ABOUT THE AUTHO

THOMAS L. BOULLION
...essor of Mathematics and
...ciate Profe... Texas Tech University
Statistics at Texas Tech University
He has also held such positions as
Research Scientist at Tracor, and
Consulting Mathematical Statistician
at the Texas Center for Research,
and Assistant Professor of
Mathematics at the University of
Southwestern Louisiana. He is the
author of numerous journal articles,
technical reports, and scientific
papers. He has co-edited one book
(Symposium Proceedings: Theory
and Applications of Generalized
Inverses of Matrices) and has
co-authored a second (Statistical
Techniques and Data Analysis).

PATRICK L. ODELL is Professor
and Chairman of the Department
of Mathematics at Texas Tech
University. He has held various
positions in the scientific community
such as Chief, Doppler Reduction
Section, White Sands Missile Range,
Meteorologist, Research Scientist,
and Consulting Mathematician to
many corporations. Dr. Odell was
elected a Fellow of the Texas
Academy of Science and also a
Fellow of the American Statistical
Association. He has been awarded
many Government Contracts to study
such topics as Optimum-Methods
For Predicting Spacecraft Vehicle
Trajectories and Computing Techniques of a Trajectory of a Re-Entry
Body. He has written numerous
monographs, journal articles, and
scientific papers. He has co-
authored three other books (Probability for Practicing Engineers, The
Generation of Random Variables,
and Estimation in Linear Models)
and was co-editor of Symposium
Proceedings: Theory and Applications of Generalized Inverses of
Matrices.

Generalized Inverse Matrices

Generalized Inverse Matrices

THOMAS L. BOULLION

Associate Professor of Mathematics and Statistics

PATRICK L. ODELL

Professor of Mathematics and Statistics

Texas Tech University, Lubbock

WILEY-INTERSCIENCE

a Division of John Wiley & Sons, Inc.

New York • London • Sydney • Toronto

Library of Congress Catalog Card Number: 79-149768

ISBN 0 471 09110 3

Printed in the United States of America

10 9 8 7 6 5 4 3 2 1

To Nancy and Norma

PREFACE

The general purpose of this book is not only to act as a source reference for researchers but also to be a text for a one-quarter or a one-semester course in advanced matrix theory for seniors or beginning graduate students. The main objective is to introduce the reader to the theory and applications of generalized inverses of matrices. In this relatively brief book, we have had to compromise with completeness and to leave untouched certain areas that could not have been neglected in a more encyclopedic presentation. However, the extensive bibliography at the end of the book should prove a valuable aid for the reader interested in pursuing special aspects of the subject matter.

Historical notes and comments as well as formal acknowledgments are collected in Appendix 2. References there and in the text are made to papers listed in the bibliography. If no numeral appears, there is only one paper or book listed under the name or names mentioned.

This text assumes that the reader has had at least a one-semester course in matrix theory or linear algebra.

At the end of each chapter a collection of problems for solution is listed. Some of these problems emphasize theoretical points; others develop an identity used in the text. We consider the problems to be important, and anyone studying this work should, at the very least, read all the problems.

Below is a brief sketch of the contents of the chapters.

Chapter 1 contains the basic definitions of various types of generalized inverses. Each type is characterized, and some fundamental properties of each are listed and discussed. The reader is encouraged to extend the list of properties. Included are some basic concepts of linear operator theory which are briefly examined.

Chapter 2 is devoted to expressions for the pseudoinverse of products, partitioned matrices, and sums of matrices as functions of the pseudoinverse of the matrices and/or their partitions.

Chapter 3 considers generalized inverses which possess specified spectral properties.

Chapter 4 deals with special topics. Range-hermitian matrices and partial isometries are characterized in terms of generalized inverses. Basic properties and extensions of well-known results are developed. The existence of differentiable generalized inverses is established. Relationships between the derivative of a matrix and that of its generalized inverse are obtained.

Chapter 5 deals with solving systems of linear equations. The concept of $p - q$ generalized inverse is introduced and utilized to obtain a best approximate solution with respect to various convex norms. Necessary and sufficient conditions for N matrix equations to have a common solution are given.

In Chapter 6 several applications of generalized inverses to statistical and nonstatistical problems are described and investigated. An algorithm for sequentially obtaining estimates of a parameter is developed. Application to distribution theory, independence of quadratic forms, finite Markov chains, and stochastic matrices are also introduced. The algebraic eigenvalue problem is solved using spectral inverses. An incidence matrix is introduced and some properties of the pseudoinverse of an incidence matrix are developed.

In the appendices the reader is presented with two techniques for calculating the pseudoinverse of a matrix. The algorithms are sufficient for the purpose of enabling the students to solve the exercises.

It is true that computer programs and error analysis of these techniques are available, but we have chosen not to include these, since our main purpose deals with the theory associated with the generalized inverse.

We give special acknowledgments with sincere appreciation to the following friends and colleagues: Professor James Scroggs, University of Arkansas; C. P. Barton, University of Texas at Austin; Professor Truman Lewis, Texas Tech University; Professor Henry P. Decell, University of Houston; Eugene Davis, Al Fieveson, and Dr. Fred Michael Speed, all of NASA—Manned Spacecraft Center, Houston. Without their interest and knowledge this book could never have been written.

We also acknowledge the financial assistance made available by the Computation and Analysis Branch of NASA—MSC through Contract NAS 9-2619 for research on the theory of generalized inverses of matrices which led to our interest in this very important topic in applied mathematics.

THOMAS L. BOULLION
PATRICK L. ODELL

Lubbock, Texas
January, 1971

CONTENTS

Generalized Inverse Matrices

I

DEFINITIONS AND FUNDAMENTAL PROPERTIES

1. DEFINITIONS

All matrices in this text are defined over the complex number field. Clearly, analogous results can be obtained by assuming the matrices are defined over the real number field.

The following matrix equations are used to define generalized inverses of matrices where ( )* denotes the conjugate transpose of the matrix:

$$AXA = A, \tag{1}$$
$$XAX = X, \tag{2}$$
$$(XA)^* = XA, \tag{3}$$
$$(AX)^* = AX. \tag{4}$$

DEFINITION 1. A generalized inverse of a matrix A is a matrix $X = A^g$ satisfying (1).

DEFINITION 2. A reflexive generalized inverse of a matrix A is a matrix $X = A^r$ satisfying (1) and (2).

DEFINITION 3. A left weak generalized inverse of a matrix A is a matrix $X = A^w$ satisfying (1), (2), and (3).

DEFINITION 3'. A right weak generalized inverse of a matrix A is a matrix $X = A^n$ satisfying (1), (2), and (4).

DEFINITION 4. A pseudoinverse of a matrix A is a matrix $X = A^+$ satisfying (1) through (4).

1

Let α^g, α^r, α^w, α^n, and α^+ denote the sets of generalized inverses, reflexive generalized inverses, left weak generalized inverses, right weak generalized inverses, and pseudoinverses of a matrix A, respectively.

The following inclusion relationships follow immediately from the above definitions:

$$\alpha^+ \subset \alpha^w \subset \alpha^r \subset \alpha^g, \tag{5}$$
$$\alpha^+ \subset \alpha^n \subset \alpha^r.$$

In view of these inclusion relationships one need only establish the existence of a pseudoinverse of a matrix which in turn establishes the existence of the remaining types of generalized inverses. However, for more insight into the relationships that exist among the various generalized inverses, we will establish the existence of a generalized inverse and show how the others can be constructed from A^g.

Theorem 1. For any matrix A there exists a generalized inverse A^g.

Proof. There exist nonsingular matrices P and Q such that

$$PAQ = \begin{pmatrix} I_r & 0 \\ 0 & 0 \end{pmatrix}$$

where r is the rank of A. Letting

$$A^g = Q \begin{pmatrix} I_r & U \\ V & W \end{pmatrix} P,$$

where U, V, and W are arbitrary, it is easily verified that $AA^gA = A$.

Theorem 2. The matrix $A^r = A^{g_1}AA^{g_2}$ is a reflexive generalized inverse of A.

Proof. This is readily established by substituting the expression for A^r in (1) and (2) and using the fact that A^{g_i} satisfies (1).

The identity $(BAA^* - CAA^*)(B - C)^* = (BA - CA)(BA - CA)^*$ along with the fact that $AA^* = 0$ implies $A = 0$ gives us the following result:

$$\text{if } BAA^* = CAA^*, \quad \text{then} \quad BA = CA. \tag{6}$$

Interchanging A and A^* we also find that

$$\text{if } BA^*A = CA^*A, \quad \text{then} \quad BA^* = CA^*. \tag{7}$$

The application of (6) and (7) to the first equation defining $(AA^*)^r$ and $(A^*A)^r$, and the conjugate transposes of these equations yield

$$A = AA^*(AA^*)^rA \tag{8}$$
$$= AA^*[(AA^*)^r]^*A \tag{9}$$
$$= A(A^*A)^rA^*A \tag{10}$$
$$= A[(A^*A)^r]^*A^*A. \tag{11}$$

Theorem 3. The matrix $A^w = A^*(AA^*)^r$ is a left weak generalized inverse of A, and $A^n = (A^*A)^rA^*$ is a right weak generalized inverse of A.

Proof. A^w satisfies (1) since, by (8), $AA^wA = AA^*(AA^*)^rA = A$. That (2) is satisfied follows from (2) in defining $(AA^*)^r$.

$$A^wAA^w = A^*(AA^*)^rAA^*(AA^*)^r = A^*(AA^*)^r = A^w.$$

Finally, from (9),

$$A^wA = A^*(AA^*)^rA = A^*(AA^*)^rAA^*[(AA^*)^r]^*A = A^*(AA^*)^rA[A^*(AA^*)^rA]^*,$$

and it follows that A^wA is hermitian. It can be established in a similar manner that $A^n = (A^*A)^rA^*$ is a right weak generalized inverse of the matrix A.

Theorem 4. The matrix $A^+ = A^wAA^n$ is a pseudoinverse of A.

Proof. From the first two defining equations for A^n and A^w we have $AA^+A = AA^wAA^nA = A$ and

$$A^+AA^+ = A^wAA^nAA^wAA^n = A^wAA^wAA^n = A^wAA^n = A^+.$$

Also, $A^+A = A^wAA^nA = A^wA$ and $AA^+ = AA^wAA^n = AA^n$. That equations (3) and (4) for A^+ are also satisfied follows from A^wA and AA^n being hermitian.

Theorems 1 to 4 provide proofs of the existence of A^g, A^r, A^w, A^n, and A^+. In general, these are not necessarily unique except for A^+. The uniqueness of A^+ will now be established.

Theorem 5. The pseudoinverse A^+ of a matrix A is unique.

Proof. To establish the uniqueness of A^+ we show that if X is any matrix satisfying the defining equations for A^+, then

$$XAA^* = A^* \tag{12}$$

and

$$X^* = AB_1 \quad \text{for some matrix } B_1. \tag{13}$$

Equation (12) is immediate, combining the conjugate transpose of (1) and (3). The conjugate transpose of (2) using (4) gives $X^* = X^*A^*X^* = A(XX^*) = AB_1$ for $B_1 = XX^*$. Likewise, A^+ satisfies (12) and (13) for some matrix, for example, B_2. Then using (12),

$$(X - A^+)AA^* = 0 \quad \text{or} \quad (X - A^+)A = 0. \tag{14}$$

Using (13), $(X - A^+) = (B_1^* - B_2^*)A^*$. This implies

$$(X - A^+)C = 0 \tag{15}$$

where the matrix C has columns orthogonal to the columns of A. Equations

(14) and (15) together imply that $X - A^+ = 0$, which establishes the uniqueness of A^+.

Other formulations that are equivalent to the above definitions of generalized inverses of a matrix are given in the exercises at the end of the chapter.

We extend now the definition of a pseudoinverse to a weighted pseudoinverse. Let R and S be given positive definite hermitian matrices of orders m and n, respectively, and let A and X satisfy the four defining equations for the pseudoinverse. Defining $Y = S^{1/2} X R^{-1/2}$ and $B = R^{1/2} A S^{-1/2}$, it is readily verified that

$$YBY = Y, \tag{16}$$

$$BYB = B, \tag{17}$$

$$(YB)^* = S^{-1} YBS, \tag{18}$$

$$(BY)^* = R^{-1} BYR. \tag{19}$$

The unique matrix Y, satisfying the above four equations for given R, S, will be called a weighted pseudoinverse and will be denoted by $Y = B^\#$. If $R = S = I$, then $B^\# = B^+$. If $S = I$, then for any given positive definite hermitian matrix R the equations (16) through (19) determine a unique left weak generalized inverse of B. Likewise, if $R = I$ and S is some given positive definite hermitian matrix, the equations (16) through (19) determine a unique right weak generalized inverse of B.

Other formulations of generalized inverses that have spectral properties not generally enjoyed by any of the above are treated separately in Chapter 3.

2. CHARACTERIZATIONS

It is well known that any rectangular matrix of rank r is equivalent to a matrix of the form

$$\begin{pmatrix} B & 0 \\ 0 & 0 \end{pmatrix}$$

where B is an $r \times r$ nonsingular matrix. Before proceeding to characterize each type of generalized inverse we establish several lemmas needed in the subsequent characterizations.

Lemma 1. A necessary and sufficient condition for the equation $AXB = C$ to have a solution is

$$AA^+ CB^+ B = C, \tag{20}$$

in which case the general solution is

$$X = A^+ CB^+ + Y - A^+ AYBB^+ \tag{21}$$

where Y is arbitrary.

Proof. If X satisfies $AXB = C$, then

$$C = AXB = AA^+AXBB^+B = AA^+CB^+B.$$

Conversely, if (20) holds then A^+CB^+ is a particular solution of $AXB = C$. Now any expression of the form $X = Y - A^+AYBB^+$ satisfies $AXB = 0$, and conversely, if $AXB = 0$, then $X = X - A^+AXBB^+$. It follows that the general solution is as given in (21).

Linear systems are discussed in more detail in a later chapter. Here we are establishing only that which will be needed to characterize generalized inverses. The characterization is based on the work of Morris and Odell [1].

Suppose A is an $m \times n$ matrix of rank r. Also, suppose

$$P = \begin{pmatrix} P_1 \\ P_2 \end{pmatrix}, \quad P^{-1} = (S_1, S_2), \quad Q = (Q_1, Q_2), \quad Q^{-1} = \begin{pmatrix} T_1 \\ T_2 \end{pmatrix}$$

where

$$PAQ = \begin{pmatrix} B & 0 \\ 0 & 0 \end{pmatrix}$$

and B is an $r \times r$ nonsingular matrix.

Lemma 2. (a) $P_1 X = I$ and $P_2 X = 0$ have a unique common solution

$$X = (I - P_2^+ P_2)(P_1 - P_1 P_2^+ P_2)^+. \tag{22}$$

(b) $XQ_1 = I$ and $XQ_2 = 0$ have a unique common solution

$$X = (Q_1 - Q_2 Q_2^+ Q_1)^+ (I - Q_2 Q_2^+).$$

Proof. Let $P^{-1} = (X, Y)$. Then

$$I = PP^{-1} = \begin{pmatrix} P_1 X & P_1 Y \\ P_2 X & P_2 Y \end{pmatrix}$$

implies that $P_1 X = I$ and $P_2 X = 0$ have a unique common solution. Obviously, the expression given for X satisfies the equations. Part (b) follows in a similar manner.

Lemma 3. (a) $S_1(P_1 + XP_2)$ is hermitian if and only if $X = -P_1 P_2^+$.

(b) $(Q_1 + Q_2 X)T_1$ is hermitian if and only if $X = -Q_2^+ Q_1$.

Proof. Assume $S_1(P_1 + XP_2)$ is hermitian. This implies $S_1(P_1 + XP_2) = (P_1 + XP_2)^* S_1^*$. Premultiplying this equation by P_1 and postmultiplying by P_2^* we have

$$P_1 S_1(P_1 + XP_2)P_2^* = P_1(P_1 + XP_2)^*(P_2 S_1)^*.$$

But $P_1 S_1 = I$ and $P_2 S_1 = 0$ so that $(P_1 + XP_2)P_2^* = 0$. Thus

$$X = -P_1 P_2^*(P_2 P_2^*)^{-1} = -P^1 P_2^+.$$

The fact that $P_2^+ = P_2^*(P_2 P_2^*)^{-1}$ follows from the fact that P_2 has full row rank. See problem 5(c) in the exercises. If $X = -P_1 P_2^+$, then

$$[S_1(P_1 + XP_2)] = [S_1 P_1(I - P_2^+ P_2)]$$
$$= (I - P_2^+ P_2)(P_1 - P_1 P_2^+ P_2)^+ (P_1 - P_1 P_2^+ P_2)$$

since $S_1 = (I - P_2^+ P_2)(P_1 - P_1 P_2^+ P_2)^+$ from Lemma 2. Now,

$$[S_1(P_1 + XP_2)]^* = (P_1 - P_1 P_2^+ P_2)^+ P_1(I - P_2^+ P_2)(I - P_2^+ P_2)$$
$$= (P_1 - P_1 P_2^+ P_2)^+ (P_1 - P_1 P_2^+ P_2)$$

since $I - P_2^+ P_2$ is idempotent. It follows that $S_1(P_1 + XP_2)$ is hermitian. Part (b) follows in a similar manner.

The characterization of generalized inverses with respect to the equivalence relation will now be established.

Theorem 6. Suppose A is $m \times n$ and

$$PAQ = \begin{pmatrix} B & 0 \\ 0 & 0 \end{pmatrix}$$

where B is an $r \times r$ nonsingular matrix. Let X be an $n \times m$ matrix and define

$$Q^{-1}XP^{-1} = \begin{pmatrix} Z & U \\ V & W \end{pmatrix}.$$

Then
 (a) X is a generalized inverse of A if and only if $Z = B^{-1}$.
 (b) X is a reflexive generalized inverse of A if and only if $Z = B^{-1}$ and $W = VBU$.
 (c) X is a right weak generalized inverse of A if and only if $Z = B^{-1}$, $U = -B^{-1}P_1 P_2^+$, and $W = -VP_1 P_2^+$.
 (d) X is a left weak generalized inverse of A if and only if $Z = B^{-1}$, $V = -Q_2^+ Q_1 B^{-1}$, and $W = -Q_2^+ Q_1 U$.
 (e) X is the pseudoinverse of A if and only if $Z = B^{-1}$, $U = -B^{-1}P_1 P_2^+$, $V = -Q_2^+ Q_1 B^{-1}$, and $W = Q_2^+ Q_1 B^{-1} P_1 P_2^+$.

Proof. Note that

$$A = P^{-1}\begin{pmatrix} B & 0 \\ 0 & 0 \end{pmatrix}Q^{-1} \quad \text{and} \quad X = Q\begin{pmatrix} Z & U \\ V & W \end{pmatrix}P.$$

 (a) $AXA = A$ if and only if $BZB = B$ or $Z = B^{-1}$.
 (b) Assume $Z = B^{-1}$. $XAX = X$ if and only if $W = VBU$.
 (c) Assume $Z = B^{-1}$ and $W = VBU$. Then

$$AX = P^{-1}\begin{pmatrix} I & BU \\ 0 & 0 \end{pmatrix}P = (S_1, S_2)\begin{pmatrix} I & BU \\ 0 & 0 \end{pmatrix}\begin{pmatrix} P_1 \\ P_2 \end{pmatrix} = S_1(P_1 + BUP_2)$$

is hermitian if and only if $BU = -P_1 P_2^+$ by Lemma 3(a). Thus $U = -B^{-1} P_1 P_2^+$.

(d) Assume $Z = B^{-1}$ and $W = VBU$. Then

$$XA = Q \begin{pmatrix} I & 0 \\ VB & 0 \end{pmatrix} Q^{-1} = (Q_1, Q_2) \begin{pmatrix} I & 0 \\ VB & 0 \end{pmatrix} \begin{pmatrix} T_1 \\ T_2 \end{pmatrix} = (Q_1 + Q_2 VB) T_1$$

is hermitian if and only if $VB = -Q_2^+ Q_1$ by Lemma 3(b). Thus $V = -Q_2^+ Q_1 B^{-1}$.

(e) Follows immediately from (a) to (d).

A characterization of the weighted pseudoinverse is given as an exercise at the end of the chapter. We now proceed to establish some basic properties of generalized inverses.

3. PROPERTIES

Several basic results that hold for any generalized inverse of a matrix A are given in the following theorem.

Theorem 7. The following are true for any generalized inverse A^g of A:
  (a) Rank $A^g \geq$ rank A.
  (b) One choice of $(A^*)^g$ is $(A^g)^*$.
  (c) $A(A^*A)^g A^* A = A$.
  (d) $A^* A(A^*A)^g A^* = A^*$.
  (e) If A is idempotent, $A^g = A$ is one choice for A^g.
  (f) $A(A^*A)^g A^*$ is a hermitian, idempotent matrix projecting vectors onto the subspace spanned by the columns of A.
  (g) $A^g A$ is idempotent.

Proof. Part (a) is immediate from rank considerations. Parts (b), (e), and (g) follow from the definition. Part (c) is established from

$$[A(A^*A)^g A^* A - A]^* [A(A^*A)^g A^* A - A]$$
$$= [(A^*A)^g A^* A - I]^* A^* [A(A^*A)^g A^* A - A] = 0$$

since $A^* A(A^*A)^g A^* A - A^* A = 0$. Part (d) is proved in a similar manner. To establish (f), consider the product of $A(A^*A)^g A^* - A[(A^*A)^g]^* A^*$ by its conjugate transpose. Using (d), the product is found to be zero and thus $A(A^*A)^g A^* = A[(A^*A)^g]^* A^*$. The other results in (f) are easily established.

Theorem 8. The following are true for any reflexive generalized inverse of A:
  (a) Parts (b) through (g) of Theorem 7, replacing A^g with A^r.
  (b) A^g is a reflexive generalized inverse of A if and only if rank $A^g =$ rank A.

Proof. Part (a) follows since $\alpha^r \subset \alpha^g$. The "only if" part of (b) is readily established from the definition of A^r. Conversely, assume $AA^gA = A$ and rank $A = $ rank A^g, then rank $A^gA = $ rank A^g. It follows that $A^g = A^gAX$ for some X, hence that $AA^g = AA^gAX = AX$. Hence $A^g = A^gAX = A^gAA^g$.

Theorem 9. The following are true for any left weak generalized inverse of A:

(a) Parts (c), (d), (f), and (g) of Theorem 7, replacing A^g with A^w.

(b) One choice for A^w is $A^*(AA^*)^g$ and for $(A^*)^w$ is $(A^n)^*$.

(c) One choice for $(A^*A)^w$ is $A^w(A^*)^w$.

(d) One choice for $(UAV)^w$ is $V^*A^wU^*$ where U and V are unitary matrices.

(e) $A^+A = A^wA$ for any choice of A^w.

(f) If B_1 and B_2 are any two choices of A^w, then $(B_1 - B_2)A = 0$.

(g) If rank $A = n = \min(m, n)$, then every A^g is a A^w.

(h) If A is hermitian and idempotent, one choice for A^w is A.

Proof. Part (a) follows since $\alpha^w \subset \alpha^r$. Parts (b) to (d), (g), and (h) follow readily from the definition. To establish (e), consider $A(A^wA)^* = AA^wA = A = AA^+A = A(A^+A)^*$. This implies $AA^*(A^w)^* = AA^*(A^+)^*$ and thus $(A^wA)^* = (A^+A)^*$ or $A^wA = A^+A$. Part (f) is immediate from (e).

Theorem 10. The following are true for any right weak generalized inverse of A:

(a) Parts (c), (d), (f), and (g) of Theorem 7, replacing A^g with A^n.

(b) One choice for A^n is $(A^*A)^gA^*$ and for $(A^*)^n$ is $(A^w)^*$.

(c) One choice for $(AA^*)^n$ is $(A^*)^nA^n$.

(d) One choice for $(UAV)^n$ is $V^*A^nU^*$ where U and V are unitary matrices.

(e) $AA^+ = AA^n$ for any choice of A^n.

(f) If B_1 and B_2 are any two choices of A^n, then $A(B_1 - B_2) = 0$.

(g) If rank $A = m = \min(m, n)$, then every A^g is a A^n.

(h) If A is a hermitian and idempotent, one choice for A^n is A.

Proof. The proof is similar to Theorem 9 and is left as an exercise.

Theorem 11. The following are true for the pseudoinverse A^+ of A:

(a) $(A^+)^+ = A$.

(b) $(A^*)^+ = (A^+)^*$.

(c) $(AA^*)^+ = (A^+)^*A^+$ and $(A^*A)^+ = A^+(A^*)^+$.

(d) rank $A = $ rank $A^+ = $ rank $A^+A = $ trace A^+A.

(e) $(UAV)^+ = V^*A^+U^*$ when U, V are unitary.

(f) $(\lambda A)^+ = \lambda^{-1}A^+$ provided $\lambda \neq 0$; $0^+ = 0^T$.

(g) $A^+ = (A^*A)^+A^*$.

(h) $A^+ = (A^*A)^wA^*$ for any choice of $(A^*A)^w$.

(i) $A^+ = A^*(AA^*)^n$ for any choice of $(AA^*)^n$.

(j) $A^+ = A^w$ if rank $A = m = \min(m, n)$.

(k) $A^+ = A^n$ if rank $A = n = \min(m, n)$.

(l) $A^+ = A^n = A^w = A^r = A^g = A^{-1}$ if A is nonsingular.

(m) $AA^+, A^+A, I - A^+A, I - AA^+$ are hermitian and idempotent.

(n) $(AA^*)^+ AA^* = AA^+$.

(o) $A^+A = AA^+$ if A is normal.

(p) $(A^k)^+ = (A^+)^k$ if A is normal and k a positive integer.

(q) $AB = 0$ if and only if $B^+A^+ = 0$.

Proof. The proof is similar to Theorem 10 and is left as an exercise.

4. OPERATOR THEORY

Many times it is advantageous to consider matrices as representations of linear operators on finite dimensional vector spaces. Since every finite dimensional inner-product vector space is a Hilbert space, the setting is assumed to be such a space. For practical purposes one could assume the setting is the Euclidean n-dimensional vector space over the complex number field denoted by E^n. A brief discussion of some results from linear algebra, which will be needed in the sequel, is in order. The linear manifold generated by the columns of a matrix is represented by $R(A)$ and will be referred to as the range space of A. The set of all column vectors X such that $AX = 0$ is the null space of A and is denoted by $N(A)$. If U and V are subspaces of E^n, the following three conditions are equivalent:

(1) $E^n = U \oplus V$.

(2) $U \cap V = 0$ and $U + V = E^n$.

(3) Every vector W in E^n may be written in the form $W = X + Y$, with X in U and Y in V, in one and only one way.

Statement (1) says that E^n is the direct sum of the subspaces U and V. Statement (2) tells us that U and V are complementary subspaces of E^n. This is denoted by writing $U = V^c$. If, in addition, all the vectors in U are orthogonal to all vectors in V, U and V are orthogonal complements of each other, and we denote this by $U^\perp = V$, or equivalently, $V^\perp = U$. For instance, $R(A)^\perp$ is the set containing the maximal number of linearly independent vectors orthogonal to the columns of the matrix A.

Especially important for our purposes is the following additional connection between direct sums and linear operators.

DEFINITION 5. If $E^n = U \oplus V$ so that every W in E^n may be written, uniquely, in the form $W = X + Y$, with X in U and Y in V, the projection on U along V is the linear operator P defined by $PW = X$.

The following properties of projections are readily established and can be found in many elementary linear algebra texts:

1. A linear operator P is a projection on some subspace if and only if it is

idempotent. It is an orthogonal projection if and only if it is both hermitian and idempotent.

2. A linear operator P is a projection if and only if $I - P$ is a projection. Moreover, if P is a projection on U along V, then $I - P$ is a projection on V along U.

For any $m \times n$ matrix A the transformation $Y = AX$ defined for all vectors X in E^n is a linear transformation on E^n to E^m.

Theorem 12. Let A be a linear operator on E^n to E^m. If $AXA = A$, then $R(A) \oplus N(AX) = E^m$ and $N(A) \oplus R(XA) = E^n$.

Proof. From $AXA = A$ we find that $(AX)^2 = AX$ so that AX is a projection in E^m. Certainly, it must be a projection on some subspace of $R(A)$. In fact, it must be onto $R(A)$, for if Y is a vector in $R(A)$, then $Y = AW$ for some vector W in E^n so that $Y = AX(AW)$. It follows that $N(AX)$ is such that $R(A) \oplus N(AX) = E^m$ and AX is a projection on $R(A)$ along $N(AX)$. Also, from $AXA = A$ we have $(XA)^2 = XA$ so that XA is a projection in E^n. The null space of XA certainly contains $N(A)$. If Y is a vector in E^n such that $XAY = 0$, then $AY = AXAY = 0$ so that Y is in $N(A)$. Hence $N(XA) = N(A)$, and it follows that $N(A) \oplus R(XA) = E^n$ with XA a projection on $R(XA)$ along $N(A)$.

Hence, if we also require that $XAX = X$, or equivalently, that the rank of X be that of A, then $R(XA) = R(X)$ and $N(AX) = N(X)$. If we further require $(XA)^* = XA$, then XA is an orthogonal projection on $R(X)$ so that $R(X)$ is the orthogonal complement of $N(A)$. Likewise, with $AXA = A$, $XAX = X$, and $(AX)^* = AX$ holding, $N(X)$ is the orthogonal complement of $R(A)$. Thus for $X = A^+$ we find that AA^+ is an orthogonal projection on $R(A)$, and A^+A is an orthogonal projection on $N(A)^\perp$.

EXERCISES

1. Given the matrix $A = \begin{bmatrix} 1 & 0 & 1 & 1 \\ 0 & 1 & -1 & 0 \\ 1 & 1 & 0 & 1 \end{bmatrix}$, calculate a member of α^g, α^r, α^w, α^n, and α^+.

2. (a) Let Y be any vector for which $AX = Y$ is a consistent equation, and define A^g so that $X = A^gY$ is a solution for all such Y. Prove that this definition of A^g is equivalent to Definition 1.

   (b) Prove that $A^*AA^gA = A^*A$ is an equivalent formulation for A^g.

3. Prove that the conditions $ABA = A$, $B = A^*C$ for some matrix C are equivalent to Definition 3 for A^w.

4. Prove that the conditions $ABA = A$, $B = CA^*$ for some matrix C are equivalent to Definition 3' for A^n.

5. Let $A = BC$ where B is $m \times r$ of rank $r \neq 0$, C is $r \times n$ of rank r. Prove:
   (a) $X = A^+$ where $X = C^*(CC^*)^{-1}(B^*B)^{-1}B^*$.
   (b) If $A = B$, $A^+ = (B^*B)^{-1}B^*$.
   (c) If $A = C$, $A^+ = C^*(CC^*)^{-1}$.
6. Prove that the following sets of conditions on B are each equivalent to Definition 4 for A^+.
   (a) $A^*AB = A^*$, $B^* = CA$ for some matrix C.
   (b) $BAA^* = A^*$, $B^* = AC$ for some matrix C.
   (c) $BAA^* = A^*$, $ABB^* = B^*$.
   (d) $A^*AB = A^*$, $B^*BA = B^*$.
7. Let A be the matrix given in problem (1),
$$R^{1/2} = \begin{pmatrix} 3 & 1 & 1 \\ 1 & 4 & 2 \\ 1 & 2 & 2 \end{pmatrix}, \quad S^{1/2} = \begin{pmatrix} R^{1/2} & 0 \\ 0 & 4 \end{pmatrix}, \quad \text{and} \quad B = R^{1/2}AS^{-1/2}. \quad \text{Calculate} \quad B^{\#}.$$

Also calculate the unique left and right weak generalized inverses of B, determined by the positive definite matrix pairs $S = I$, R and S, $R = I$, respectively.

8. Let $A = P^{-1}\begin{pmatrix} B & 0 \\ 0 & 0 \end{pmatrix}Q^{-1}$, $X = Q\begin{pmatrix} Z & U \\ V & W \end{pmatrix}P$ where B is $r \times r$ and nonsingular.
   (a) Prove that $X = A^{\#}$ if and only if $Z = B^{-1}$, $U = -B^{-1}P_1RP_2^*(P_2RP_2^*)^{-1}$,
   $V = -(Q_2^*SQ_2)^{-1}Q_2^*SQ_1B^{-1}$, $W = VBU$, where $P = \begin{pmatrix} P_1 \\ P_2 \end{pmatrix}$, $P^{-1} = (S_1, S_2)$,
   $Q = (Q_1, Q_2)$, $Q^{-1} = \begin{pmatrix} T_1 \\ T_2 \end{pmatrix}$.
   (b) Show that the characterization of A^+ given in Theorem 6 follows from the result in (a).
9. Prove Theorem 10.
10. Prove Theorem 11.
11. Show that if A is a matrix defined on the real field, the orthogonal complement of the null space of A is the range space of A^T, that is, $N(A)^{\perp} = R(A^T)$.

2

PSEUDOINVERSES OF SUMS AND PRODUCTS

1. THE PSEUDOINVERSE OF PRODUCTS

It will be assumed throughout that the sizes of the matrices are conformable to the operations of addition and multiplication. For square nonsingular matrices it is always true that $(AB)^{-1} = B^{-1}A^{-1}$. However,

$$(AB)^+ = B^+A^+ \tag{1}$$

does not hold in general, as can be seen from the following example:

$$A = (1, 0), \quad B = (1, 1)^T \quad (AB)^+ = (1), \quad B^+A^+ = (1/2, 1/2)\begin{pmatrix}1\\0\end{pmatrix} = (\tfrac{1}{2}).$$

If A is of full column rank and B of full row rank, it is readily established that (1) holds since in that case

$$(AB)^+ = B^*(BB^*)^{-1}(A^*A)^{-1}A^* = B^+A^+.$$

The general expression for $(AB)^+$ is given in the next theorem.

Theorem 1. Let $B_1 = A^+AB$ and $A_1 = AB_1B_1^+$, then $(AB)^+ = B_1^+A_1^+$.

Proof. The product AB can be written as

$$AB = AA^+AB = AB_1 = AB_1B_1^+B_1 = A_1B_1.$$

Letting $X = B_1^+A_1^+$ and $Y = AB$, it is only necessary to show that X and Y satisfy the defining equations for the pseudoinverse. Since $A_1 = AB_1B_1^+ = AB_1B_1^+B_1B_1^+ = A_1B_1B_1^+$, it follows that $YX = A_1B_1B_1^+A_1^+ = A_1A_1^+$ is hermitian. Also, $YXY = A_1B_1B_1^+A_1^+A_1B_1 = A_1A_1^+A_1B_1 = A_1B_1 = Y$ and $XYX = B_1^+A_1^+A_1B_1B_1^+A_1^+ = B_1^+A_1^+A_1A_1^+ = B_1^+A_1^+ = X$.

In order to show XY is hermitian we observe first that

$$A^+A_1 = A^+AB_1B_1^+ = A^+A(A^+AB)B_1^+ = B_1B_1^+.$$

Also, $A_1^+A_1 = A_1^+AB_1B_1^+ = A_1^+A_1B_1B_1^+$. Taking the conjugate transpose of this we have $A_1^+A_1 = B_1B_1^+A_1^+A_1 = A^+A_1A_1^+A_1 = A^+A_1 = B_1B_1^+$. Thus $XY = B_1^+A_1^+A_1B_1 = B_1^+B_1B_1^+B_1$ is hermitian and the conclusion follows.

A necessary and sufficient condition for (1) to hold is given next.

Theorem 2. $(AB)^+ = B^+A^+$ if and only if both the equations,

$$A^+AB(AB)^* = B(AB)^* \tag{2}$$

and

$$BB^+A^*AB = A^*AB, \tag{3}$$

are satisfied.

Proof. Pre- and postmultiplying (2) by B^+ and $(AB)^{*+}$, respectively, yield $B^+A^+AB(AB)^*(AB)^{*+} = B^+B(AB)^*(AB)^{*+}$. Using the property $C^+CC^* = C^*$ to simplify gives $B^+A^+AB = (AB)^*(AB)^{*+} = (AB)^+(AB)$. Similarly, pre- and postmultiplying the conjugate transpose of (3) by $(AB)^{*+}$ and A^+, respectively, and simplifying give $ABB^+A^+ = (AB)(AB)^+$. Thus $(AB)^+ = B^+A^+$ because of the uniqueness of $(AB)^+$ in the orthogonal projections $AB(AB)^+$ and $(AB)^+AB$, whenever $R[(AB)^+] \subset R[(AB)^*]$ and $R[(AB)^{+*}] \subset R(AB)$.

Conversely, $(AB)^+ = B^+A^+$ implies $(AB)^* = B^+A^+(AB)(AB)^*$. Premultiplying by ABB^*B and using $B^*BB^+ = B^*$ give $ABB^*(I - A^+A)B(AB)^* = 0$.

Since the left member is hermitian and $I - A^+A$ is idempotent, it follows that $(I - A^+A)B(AB)^* = 0$, which is equivalent to (2). In a similar manner (3) is obtained.

Other equivalent forms of the above necessary and sufficient condition for (1) are given in the exercises.

Equations (2) and (3) have a simple interpretation in terms of range spaces. They assert that $R(A^*)$ is an invariant space of BB^*, and $R(B)$ is an invariant space of A^*A.

2. THE PSEUDOINVERSE OF A PARTITIONED MATRIX

Consider an arbitrary matrix $A = (U, V)$ where U and V have s and $n-s$ columns, respectively. We assume that $0 < s < n$. Before establishing the general form for A^+ we establish the following theorem which will be instrumental in establishing the more general result.

Theorem 3. Let $A = (U, V)$ and

$$X_1 = \begin{bmatrix} U^+ - U^+VC^+ - U^+V(I - C^+C)KV^*U^{+*}U^+ \\ C^+ + (I - C^+C)KV^*U^{+*}U^+ \end{bmatrix} \tag{4}$$

where $C = (I - UU^+)V$ and $K = (I + V^*U^{+*}U^+V)^{-1}$, then $X_1 = A^+$ if and only if C^+C and $V^*U^{+*}U^+V$ commute.

Proof. It will be established that X_1 satisfies the defining equations for A^+, where the commutativity of C^+C and $V^*U^{+*}U^+V$ is used to establish that X_1A is hermitian.

The product AX_1 simplifies to $AX_1 = UU^+ + CC^+$, and since UU^+ and CC^+ are hermitian, AX_1 is hermitian. From the definition of C we have $V = UU^+V + C$ and $U^+C = 0$. Thus

$$AX_1A = [(UU^+ + CC^+)U, (UU^+ + CC^+)V] = [U, V] = A$$

since $U^+C = 0$, which implies that $C^+U = 0$. Similarly, X_1AX_1 reduces to

$$X_1AX_1 = \begin{bmatrix} U^+ - U^+VC^+ - U^+V(I - C^+C)KV^*U^{+*}U^+ \\ C^+ + (I - C^+C)KV^*U^{+*}U^+ \end{bmatrix} = X_1$$

since $U^+(UU^+ + CC^+) = U^+$ and $C^+(UU^+ + CC^+) = C^+$. Finally, the product X_1A becomes

$$X_1A = \begin{bmatrix} U^+U - U^+V(I - C^+C)KV^*U^{+*} & U^+V(I - C^+C)K \\ (I - C^+C)KV^*U^{+*} & I - (I - C^+C)K \end{bmatrix} \tag{5}$$

using $C^+U = 0$, $C^+V = C^+C$, $U^{+*}U^+U = U^{+*}$, where $KV^*U^{+*}U^+V = I - K$ by definition of K. Now, if C^+C and $V^*U^{+*}U^+V$ commute, then $(I - C^+C)(I + V^*U^{+*}U^+V) = (I + V^*U^{+*}U^+V)(I - C^+C)$, and so $K(I - C^+C) = (I - C^+C)K$. Since K and $I - C^+C$ are hermitian, it follows in (5) that X_1A is hermitian. Thus $X_1 = A^+$ follows from the uniqueness of A^+.

Conversely, if $X_1 = A^+$, then $(X_1A)^* = X_1A$, and the fact that $K(I - C^+C)$ is hermitian is immediate from (5). Hence $K(I - C^+C) = (I - C^+C)K$ and C^+C commutes with $V^*U^{+*}U^+V$.

Special cases where the form of A^+ simplifies are given in the exercises. The general form for A^+ is given below.

Theorem 4. Let $A = (U, V)$, then

$$A^+ = \begin{bmatrix} U^+ - U^+VC^+ - U^+V(I - C^+C)MV^*U^{+*}U^+(I - VC^+) \\ C^+ + (I - C^+C)MV^*U^{+*}U^+(I - VC^+) \end{bmatrix}$$

where $C = (I - UU^+)V$ and $M = [I + (I - C^+C)V^*U^{+*}U^+V(I - C^+C)]^{-1}$.

Proof. Let X_0 be the matrix

$$X_0 = \begin{bmatrix} U^+ - U^+VC^+ - U^+V(I - C^+C)L \\ C^+ + (I - C^+C)L \end{bmatrix}$$

obtained from X_1 in (4) by replacing $KV^*U^{+*}U^+$ by an arbitrary matrix, L, of the same size. It is readily verified that $AX_0 = UU^+ + CC^+ = AX_1$ using the definition of C and the relation $C(I - C^+C) = 0$. Hence we have $(AX_0)^* = AX_0$ and $AX_0 A = A$ from the proof of Theorem 3. Now, $X_0(AX_0) = X_0$ provided L satisfies $L(UU^+ + CC^+) = L$. Also, $X_0 A$, after some simplification, becomes

$$X_0 A = \begin{bmatrix} U^+U - U^+V(I - C^+C)LU & U^+V(I - C^+C)(I - LV) \\ (I - C^+C)LU & C^+C + (I - C^+C)LV \end{bmatrix},$$

and it follows that $X_0 A$ is hermitian provided

$$[U^+V(I - C^+C)(I - LV)]^* = (I - C^+C)LU, \tag{6}$$

and also that $U^+V(I - C^+C)LU$ and $(I - C^+C)LV$ are hermitian. We show now that $L = MV^*U^{+*}U^+(I - VC^+)$ satisfies those conditions. Since $U^+(I - VC^+)UU^+ = U^+$ and $U^+(I - VC^+)CC^+ = -U^+VC^+$, the above expression for L satisfies $L(UU^+ + CC^+) = L$. Now, since $I - C^+C$ is idempotent, it commutes with $I + (I - C^+C)V^*U^{+*}U^+V(I - C^+C)$, and thus with M. Since $I - C^+C$ and M are hermitian, $[(I - C^+C)M]^* = (I - C^+C)M$, and so $U^+V(I - C^+C)LU = U^+V(I - C^+C)MV^*U^{+*}$ is hermitian. Also, $(I - C^+C)LV = (I - C^+C)M(I - C^+C)V^*U^{+*}U^+V(I - C^+C)$ or $(I - C^+C)LV = (I - C^+C)(I - M)$, which implies that $(I - C^+C)LV$ is hermitian. Also, since

$$U^+V(I - C^+C)(I - LV) = U^+V(I - C^+C)M = [(I - C^+C)MV^*U^{+*}]^*$$

and $(I - C^+C)LU = (I - C^+C)MV^*U^{+*}$, (6) holds for this choice of L.

It has been shown that X_0 satisfies the defining equations for A^+, provided L has the given form, so that $X_0 = A^+$.

Suppose we now consider the matrix A partitioned as $A = [U, V]$, and assume A^+ is known. Partition A^+ as

$$A^+ = \begin{bmatrix} G \\ H \end{bmatrix}$$

where G and H are of the same sizes as U^* and V^*, respectively. The next theorem establishes an expression for U^+ in terms of G, H, and related matrices.

Theorem 5. With A and A^+ partitioned as above,

$$U^+ = G[I + V(I - HV)^+H](I - CC^+)$$

where $C^+ = H - (I - HV)(I - HV)^+H$.

Proof. From the expression for A^+ in Theorem 4 we have

$$G = U^+ - U^+VC^+ - U^+V(I - C^+C)MV^*U^{+*}U^+(I - VC^+)$$

and $H = C^+ + (I - C^+C)MV^*U^{+*}U^+(I - VC^+)$, where $C = (I - UU^+)V$ and $M = [I + (I - C^+C)V^*U^{+*}U^+V(I - C^+C)]^{-1}$. Forming the product and simplifying, using the definition of L and $(I - C^+C)LV = (I - C^+C)(I - M)$ given in the proof of Theorem 4, we have

$$GV = U^+V(I - C^+C)M + U^+VC^+C + U^+(I - VC^+)CC^+V.$$

Also, using the relations $U^+(I - VC^+)CC^+ = -U^+VC^+$, $U^+C = 0$, and $C^+U = 0$, we have

$$GV = U^+V(I - C^+C)M. \tag{7}$$

Also, $HV = C^+V + (I - C^+C)(I - M)$ and it follows that

$$I - HV = (I - C^+C)M. \tag{8}$$

Since (2) and (3) hold for $A = I - C^+C$ and $B = M$, we have from Theorem 2

$$(I - HV)^+ = M^{-1}(I - C^+C) \tag{9}$$

and thus using (7),

$$GV(I - HV)^+ = U^+V(I - C^+C)MM^{-1}(I - C^+C) = U^+V(I - C^+C).$$

Combining (8) and (9), $(I - HV)(I - HV)^+ = I - C^+C$. Now,

$$GV(I - HV)^+H = U^+V(I - C^+C)H$$
$$= U^+V(I - C^+C)MV^*U^{+*}U^+(I - VC^+)$$

and so

$$G[I + V(I - HV)^+H] = U^+ - U^+VC^+. \tag{10}$$

Also,

$$(I - HV)(I - HV)^+H = (I - C^+C)MV^*U^{+*}U^+(I - VC^+),$$

hence $H - (I - HV)(I - HV)^+H = C^+$. Postmultiplying (10) by $(I - CC^+)$, since $U^+C = 0$, we have $U^+ = G[I + V(I - HV)^+H](I - CC^+)$, the desired expression.

Special cases where the form of U^+ simplifies are given in the exercises. We obtain next expressions for the pseudoinverse of certain sums of matrices.

3. THE PSEUDOINVERSE OF SUMS

Theorem 6. Let U and V be any two matrices with the same number of rows, then $(UU^* + VV^*)^+ = (I - C^{+*}V^*)U^{+*}KU^+(I - VC^+) + (CC^*)^+$ where $K = I - U^+V(I - C^+C)M(U^+V)^*$ with C and M as given in Theorem 4.

Proof. Letting $A = [U, V]$, $UU^* + VV^* = AA^*$ so that

$$(UU^* + VV^*)^+ = A^{+*}A^+.$$

The result is then obtained from Theorem 4 by block multiplication to form $A^{+*}A^+$ and then simplifying. Letting $L = (I - C^+C)MV^*U^{+*}U^+(I - VC^+)$, the expression for A^+ in Theorem 4 can be written as

$$A^+ = \begin{bmatrix} KU^+(I - VC^+) \\ C^+ + L \end{bmatrix}.$$

Noting that $L^*C^+ = 0$ we have $(C^+ + L)^*(C^+ + L) = (CC^*)^+ + L^*L$. Also, since $I - C^+C$ is idempotent and commutes with the hermitian matrix M,

$$K^*K = I - 2U^+V(I - C^+C)MV^*U^{+*}$$
$$+ U^+V(I - C^+C)M(I - C^+C)V^*U^{+*}U^+V(I - C^+C)MV^*U^{+*}$$

or

$$K^*K = I - U^+V(I - C^+C)MV^*U^{+*} - U^+V(I - C^+C)M^2V^*U^{+*},$$

where $(I - C^+C)V^*U^{+*}U^+V(I - C^+C)M = I - M$ by the definition of M. Premultiplying K^*K by $(I - C^{+*}V^*)U^{+*}$ and postmultiplying by $U^+(I - VC^+)$ and noting that the last term becomes $-L^*L$ while the others become $(I - C^{+*}V^*)U^{+*}KU^+(I - VC^+)$ yield the desired expression for $(UU^* + VV^*)^+$.

An expression for the pseudoinverse of the sum $U + V$ where U, V are arbitrary rectangular matrices such that $UV^* = 0$ is now given.

Theorem 7. Let U and V be any two matrices having the same dimensions with $UV^* = 0$, then

$$(U + V)^+ = U^+ + (I - U^+V)$$
$$\times [C^+ + (I - C^+C)MV^*U^{+*}U^+(I - VC^+)]$$

where C, M are as given in Theorem 4.

Proof. Let U, V be any two matrices with $UV^* = 0$, then

$$(U + V)(U + V)^* = UU^* + VV^*.$$

Since $A^+ = A^*(AA^*)^+$, we have

$$(U + V)^+ = (U + V)^*(UU^* + VV^*)^+$$
$$= U^*(UU^* + VV^*)^+ + V^*(UU^* + VV^*)^+. \tag{11}$$

Also, considering the partitioned matrix $A = [U, V]$, we have

$$A^+ = A^*(AA^*)^+ = \begin{bmatrix} U^*(UU^* + VV^*)^+ \\ V^*(UU^* + VV^*)^+ \end{bmatrix}. \tag{12}$$

Since A^+ is unique, corresponding submatrices in the result of Theorem 4 and (12) must be equal. Substituting these expressions for $U^*(UU^* + VV^*)^+$ and $V^*(UU^* + VV^*)^+$ in (11) yield

$$(U + V)^+ = U^+ - U^+ VC^+ - U^+ V(I - C^+C)MV^*U^{+*}U^+(I - VC^+)$$
$$+ C^+ + (I - C^+C)MV^*U^{+*}U^+(I - VC^+)$$
$$= U^+ + (I - U^+V)[C^+ + (I - C^+C)MV^*U^{+*}U^+(I - VC^+)].$$

EXERCISES

1. Prove that each of the following is both necessary and sufficient for $(AB)^+ = B^+A^+$.
   (a) A^+ABB^* and A^*ABB^+ are hermitian.
   (b) $A^+ABB^*A^*ABB^+ = BB^*A^*A$.
   (c) $A^+AB = B(AB)^+AB$ and $BB^+A^* = A^*AB(AB)^+$ are satisfied.
2. Show that A^+A and BB^+ commuting is a necessary condition for $(AB)^+ = B^+A^+$, and given an example to show that it is not sufficient.
3. Construct a numerical example for which C^+C and $V^*U^{+*}U^+V$ do not commute, thus establishing that the form of X_1 in Theorem 3 does not provide the most general form for A^+.
4. Let $A = (U, V)$, $C = (I - UU^+)V$, and $K = (I + V^*U^{+*}U^+V)^{-1}$. Prove the following corollaries to Theorem 4.
   (a) $A^+ = \begin{bmatrix} U^+ - U^+VKV^*U^{+*}U^+ \\ C^+ + KV^*U^{+*}U^+ \end{bmatrix}$ if and only if $C^+CV^*U^{+*}U^+V = 0$.
   (b) $A^+ = \begin{bmatrix} U^+ - U^+VKV^*U^{+*}U^+ \\ KV^*U^{+*}U^+ \end{bmatrix}$ if and only if $C = 0$.
   (c) $A^+ = \begin{bmatrix} U^+ - U^+VC^+ \\ C^+ \end{bmatrix}$ if and only if $C^+CV^*U^{+*}U^+V = V^*U^{+*}U^+V$.
   (d) $A^+ = \begin{bmatrix} U^+ \\ V^+ \end{bmatrix}$ if and only if $C = V$.
5. Establish that if $A = (B, C)$ where B is the submatrix of A consisting of the first $(n - 1)$ columns and C is the nth column then $A^+ = \begin{bmatrix} B^+ - B^+Cb \\ b \end{bmatrix}$ where

$$b = \begin{cases} (C - BB^+C)^+ & \text{if} \quad (I - BB^+)C \neq 0 \\ (1 + C^*(BB^*)^+C)^{-1}C^*(BB^*)^+ & \text{if} \quad (I - BB^+)C = 0. \end{cases}$$

Also, letting $d = (I - BB^+)C$, show that b can be expressed as

$$b = d^+ + (1 - d^+d)(1 + C^*(BB^*)^+C)^{-1}C^*(BB^*)^+.$$

6. Let $A = [U, V]$, $A^+ = \begin{bmatrix} G \\ H \end{bmatrix}$ where G and H are of the same sizes as U^* and V^*, respectively. Prove the following corollaries to Theorem 5 with C^+ as given in Theorem 5.

(a) $U^+ = G[I + V(I - HV)^+ H]$ if and only if $VC^+ V = C$.
(b) $U^+ = G[I - H^+ H]$ if and only if $VC^+ V = V$.
(c) $U^+ = G$ if and only if $C = V$.
In terms of G and H
(d) If $I - HV$ is nonsingular, then $U^+ = G[I + V(I - HV)^+ H]$.
(e) $U^+ = G(I - H^+ H)$ if and only if HV is idempotent.
(f) $U^+ = G$ if and only if HV is idempotent and VH is hermitian.

3

GENERALIZED INVERSES WITH SPECTRAL PROPERTIES

1. INTRODUCTION

Given a nonsingular matrix, it is well known that the eigenvalues of its inverse are the reciprocals of the eigenvalues of the matrix. In this chapter special matrices associated with a given singular square matrix that have some of the spectral properties possessed by the ordinary inverse of a nonsingular matrix are investigated.

A vector $x \neq 0$ will be called a principal vector of A of grade p corresponding to the eigenvalue λ if it satisfies the two conditions

$$(A - \lambda I)^p x = 0, \quad (A - \lambda I)^{p-1} x \neq 0 \tag{1}$$

where p is a positive integer. For brevity, x will be called a λ-vector of A of grade p. The space generated by all such vectors will be referred to as the λ-space of A. The space composed of the direct sum of all λ-spaces of A over all nonzero values of λ will be called the generalized range of A. Likewise, for row vectors $y \neq 0$ such that $y(A - \lambda I)^p = 0$, and $y(A - \lambda I)^{p-1} \neq 0$ for some positive integer p. Unless explicitly stated otherwise, all vectors in this chapter are assumed to be column vectors.

Given any square matrix A, there exists a nonsingular matrix P such that

$$A = P^{-1} J P \tag{2}$$

where J is the Jordan canonical form of A. That is, J is of the form

$$J = \begin{pmatrix} J_1 & & & \\ & J_2 & & \\ & & \ddots & \\ & & & J_k \end{pmatrix}$$

where J_i is of the form

$$J_i = \begin{pmatrix} \lambda_i & 1 & 0 & 0 \\ & \lambda_i & 1 & \\ & 0 & \ddots & 1 \\ & & & \lambda_i \end{pmatrix}$$

with the order of J_i being the maximum grade of the λ_i-vectors of A. The matrices J_i will be referred to as Jordan blocks.

2. SPECTRAL INVERSES

The two spectral properties that will be investigated are now defined.

DEFINITION 1. A generalized inverse A^g has property R if the reciprocals of nonzero eigenvalues of A are eigenvalues of A^g and conversely.

DEFINITION 2. A generalized inverse A^g has property V if whenever x is a principal vector (of grade 1) of A with nonzero eigenvalue λ, then x is a principal vector of A^g with eigenvalue λ^{-1} and conversely.

Generalized inverses do not necessarily possess properties R or V.

It is well known that any square matrix A can be written in the form

$$A = P^{-1}\begin{pmatrix} B & 0 \\ 0 & C \end{pmatrix}P \tag{3}$$

where B is nonsingular and C is nilpotent. The smallest integer k such that $C^k = 0$ is called the index of A for A singular. A nonsingular matrix will be regarded as having index 0. We first consider the situation for the class of matrices of index 1. In this case, the Jordan form of A may be written as

$$J = \begin{pmatrix} J_1 & 0 \\ 0 & 0 \end{pmatrix} \tag{4}$$

where J_1 is the direct sum of the blocks corresponding to nonzero eigenvalues. Note that the zeros or J_1 may be missing depending on the rank of A. In all cases, however, it is readily established that J commutes with its pseudo-inverse.

Theorem 1. The matrix equations

$$AXA = A, \tag{5}$$
$$XAX = X, \tag{6}$$
$$AX = XA, \tag{7}$$

have a solution if and only if A has index 0 or 1. If a solution exists, it is unique.

Proof. A has index 0 if and only if A is nonsingular and in that case $X = A^{-1}$. Suppose now that the index of A is not zero. If A has index 1, it is readily verified that $X = P^{-1}J^{+}P$ satisfies (5) to (7) with $A = P^{-1}JP$ since in that case $J^{+}J = JJ^{+}$. Assuming a solution exists, from (5) and (7) it follows that $A = AXA = A^{2}X$. Thus rank $A^{2} \geq$ rank A. But rank $A^{2} \leq$ rank A so that A and A^{2} have the same rank, which is true if and only if A has index 1.

Assume X and Y are distinct solutions to (5) through (7), then $AXAY = AY$. Also $AXAY = XAYA = XA = AX$, thus $AX = AY$. Hence $X = XAX = XAY = YAY = Y$.

From the expression $X = P^{-1}J^{+}P$, it follows that the linearly independent principal vectors are preserved and the nonzero eigenvalues are inverted. In other words, X has property V.

For the case where A may have index greater than 1, consider the following system of equations:

$$XAX = X, \tag{8}$$

$$AX = XA, \tag{9}$$

$$A^{k+1}X = A^{k}, \tag{10}$$

for some non-negative integer k.

Theorem 2. The system of equations (8) to (10) has a solution if and only if $k \geq q$ where q is the index of A. The solution, when it exists, is unique and will be referred to as the Drazin inverse.

Proof. Expressing A as in (3), if $k \geq q$, it is readily verified that

$$X = P^{-1}\begin{pmatrix} B^{-1} & 0 \\ 0 & 0 \end{pmatrix}P \tag{11}$$

is a solution. If A is the null matrix or a nonsingular matrix, the solutions are $X = 0$ and $X = A^{-1}$, respectively. In the general case, let

$$X = P^{-1}\begin{pmatrix} Z & U \\ V & W \end{pmatrix}P \tag{12}$$

be the partitioned form of a possible solution to (8) through (10), conformable to the partition

$$A = P^{-1}DP = P^{-1}\begin{pmatrix} B & 0 \\ 0 & C \end{pmatrix}P$$

with B nonsingular and C nilpotent. By introducing the partitioned forms for A and X in (8) to (10), we are led to

$$WCW = W, \tag{13}$$
$$WC = CW, \tag{14}$$
$$C^{k+1}W = C^k, \tag{15}$$

with $Z = B^{-1}$, $U = 0$, and $V = 0$. Thus X has the form

$$\begin{pmatrix} B^{-1} & 0 \\ 0 & W \end{pmatrix}$$

with W satisfying (13) to (15). If $k < q$, then $C^k \neq 0$. However, from (13) to (15) we get

$$C^k = C^{k+1}W = C^{k+1}(WCW) = C^{k+1}CW^2 = C^{k+2}W^2 = \cdots = C^{k+3}W^3 = \cdots,$$

thus

$$C^k = C^{k+s}W^s, \quad s = 1, 2, \ldots. \tag{16}$$

But when $s = q - k$ in (16) we have $C^k = C^q W^{q-k} = 0$, which is a contradiction. It follows that (8) to (10) have no solution when $k < q$, and the smallest k for which a solution exists is the index of A.

To establish uniqueness, assume X and Y are solutions to (8) through (10) with $k \geq q$ and $l \geq q$, respectively. Without loss of generality, assume $k \leq l$. Then

$$X = XAX = AXX = A(XAX)X = A^2X^3 = \cdots = A^lX^{l+1} = YA^{l+1}X^{l+1}$$
$$= YA^{l-k}(A^{k+1}X)X^l = YA^{l-k}A^kX^l = YA^lX^l = Y^2A^{l+1}X^l = \cdots = Y^{l+2}A^{l+1}$$
$$= Y^l(YAY)A^l = Y^{l+1}A^l = \cdots = Y^2A = Y.$$

Note that if k is unrestricted in (10), the system always admits a unique solution. Moreover, the expression for X in (11) is that solution.

If the index of A is one, (10) with $k = 1$ reduces to $AXA = A$ by using (9). In that case, systems (5) to (7) and (8) to (10) are the same.

It is evident from the form of X in (11) that X has property V.

When the index of A is greater than 1, the Drazin inverse is not a reflexive generalized inverse, since in that case X is of lower rank than A and does not satisfy the equation $AXA = A$.

To investigate the possibility of there being reflexive generalized inverses that satisfy the spectral properties we consider the following system of equations:

$$AXA = A, \tag{17}$$
$$XAX = X, \tag{18}$$
$$A^kX = XA^k, \tag{19}$$
$$AX^k = X^kA, \tag{20}$$

where k is some positive integer and A is given. Expressing A as in (3) and a possible solution Y for X as given in (12), and substituting these partitioned forms in (17) to (20) yield the following after some simplification:

$$Z = B^{-1},$$
$$UC = 0, \quad CV = 0,$$
$$U = B^{-k}(UC)C^{k-1} = 0,$$
$$V = C^{k-1}(CV)B^{-k} = 0.$$

Hence the solutions X, when they exist, are of the partitioned block-diagonal form:

$$X = P^{-1}\begin{pmatrix} B^{-1} & 0 \\ 0 & W \end{pmatrix}P. \tag{21}$$

The following subsystem of equations is immediate from the partitioned forms for A and X substituted in (17) to (20):

$$CWC = C, \tag{22}$$
$$WCW = W, \tag{23}$$
$$C^kW = WC^k, \tag{24}$$
$$CW^k = W^kC. \tag{25}$$

We are now in a position to establish the following result concerning the existence of solutions to the system (17) to (20).

Theorem 3. The system of equations (17) to (20) has at least one solution if and only if $k \geq q$ where q is the index of the nilpotent matrix C.

Proof. If $k < q$, then $C^k \neq 0$. The pairs of equations, (22), (24) and (23), (25), lead us to the following equations, respectively:

$$C^h = C^{h+1}W = WC^{h+1}, \tag{26}$$
$$W^h = W^{h+1}C = CW^{h+1}, \quad \text{for all } h \geq k. \tag{27}$$

From (22) and (26) we get successively:

$$C^k = C^{k-1}(CWC) = (C^kW)C = WC^{k+1}$$
$$= W^2C^{k+2} = \cdots,$$

therefore

$$C^k = W^sC^{k+s} \quad s = 1, 2, \ldots. \tag{28}$$

Similarly, from (23) and (27) we get

$$W^k = C^sW^{k+s} \quad s = 1, 2, \ldots.$$

Hence, when $s = q - k$ in (28), we have a contradiction to the fact that $C^k \neq 0$, thereby making the condition $k \geq q$ necessary.

By assuming $k \geq q$, it is readily verified that expressing A as in (3) gives

$$X = P^{-1}\begin{pmatrix} B^{-1} & 0 \\ 0 & C^r \end{pmatrix}P$$

where C^r is such that $C(C^r)^k = (C^r)^k C$ is a solution.

Certainly, if k is unrestricted, the system (17) to (20) always admits solutions. This system is invariant under similarity and it is readily deduced that the largest subspace which is invariant under A is also invariant under X. Certainly the unique solution of (17) to (20) for the nonsingular part is the ordinary inverse.

Rather than establish spectral properties for the solutions of the system (17) to (20) we establish another approach, owing to Greville [12], to the problem of obtaining a generalized inverse which has spectral properties, and show that it is equivalent to the system (17) to (20).

DEFINITION 3. A and A^s are defined to be spectral inverses if they are reflexive generalized inverses and also have the same generalized range space and the same 0-space ($\lambda = 0$).

DEFINITION 4. The core of a square matrix A is the matrix C_A defined by $C_A = (A^D)^D = AA^D A$.

DEFINITION 5. The nilpotent part of a square matrix A is the matrix N_A defined by $N_A = A - C_A = (I - AA^D)A$.

Lemma 1. The row 0-space of any square matrix A is the orthogonal complement of the column generalized range space of A.

Proof. Let x be a vector in the row 0-space of A such that $xA^k = 0$, and y be in the column generalized range of A such that $(A - \mu I)^p y = 0$, $\mu \neq 0$. Expanding $(A - \mu I)^p y$, it is readily shown that y can be expressed as

$$y = \sum_{i=1}^{p} c_i A^i y,$$

since the coefficient of y is not zero. Hence

$$xy = \sum_{i=1}^{p} c_i x A^i y. \tag{29}$$

If $k = 1$, $xy = 0$ from (29). Assume $xy = 0$ for all y in the column μ-space of A, and all x such that $xA^k = 0$ for some $k = r$. If this is all the row 0-space, we are done. If not, there exists a vector z in the row 0-space such that $zA^{r+1} = 0$. For $i = 1, \ldots, r$, zA^i is a row 0-vector of grade r or less and by the induction hypothesis, $zA^iy = 0$. If $p \le r$, from (29) $zy = 0$. If $p > r$, $zA^{r+1} = \cdots = zA^p = 0$ and again from (29) $zy = 0$. Hence, by induction, $xy = 0$ for any x in the row 0-space of A and y in the column generalized range space of A.

Theorem 4. The core C_A and A^D are spectral inverses.

Proof. By definition, $C_A = (A^D)^D$. Since the index of A^D is 0 or 1 (see problem 3), it follows from the defining equations for $(A^D)^D$ that C_A is a reflexive generalized inverse of A^D.

If A is nonsingular, certainly A and A^D are spectral inverses. Therefore we shall assume throughout the remainder of this proof that A is singular. Note that the 0-space and the null space of A^D are identical since the index of A^D is 1. (See problem 1.d.) Let x be a vector in the null space of A^D so that $A^D x = 0$. This implies $A^{k+1} A^D x = A^k x = 0$ so that x is in the 0-space of A. Similarly, the row null space of A^D is contained in the row 0-space of A. To show that the 0-space of A is contained in the null space of A^D, partition A as

$$A = P^{-1} \begin{pmatrix} B & 0 \\ 0 & C \end{pmatrix} P$$

where B is nonsingular and C is nilpotent. Let x be in the 0-space of A, then

$$0 = A^k x = P^{-1} \begin{pmatrix} B^k & 0 \\ 0 & C^k \end{pmatrix} P x = P^{-1} \begin{pmatrix} B^k & 0 \\ 0 & C^k \end{pmatrix} \begin{pmatrix} y_1 \\ y_2 \end{pmatrix}$$

so that y_1 must be null. Thus

$$A^D x = P^{-1} \begin{pmatrix} B^{-1} & 0 \\ 0 & 0 \end{pmatrix} P x = P^{-1} \begin{pmatrix} B^{-1} & 0 \\ 0 & 0 \end{pmatrix} \begin{pmatrix} 0 \\ y_2 \end{pmatrix} = 0$$

so that x is in the null space of A^D. Similarly, the row null space of A^D and the row 0-space of A are identical. Since the row 0-space of a square matrix is the orthogonal complement of the column generalized range space, it follows that A and A^D have the same generalized range space and the same 0-space. Hence, with A^D playing the role of A, it follows that A^D and $C_A = (A^D)^D$ have the same generalized range space and the same 0-space and are thus spectral inverses.

Theorem 5. $A = C_A + N_A$ is the only decomposition of A of the form $A = B + C$ such that B has index 0 or 1, C is nilpotent, and $BC = CB = 0$.

Proof. Let $A = B + C$ be such a decomposition. Then $B^D C = B^D B B^D C = B^D B^D BC = 0$. Similarly, $CB^D = 0$. Hence $AB^D = (B + C)B^D = BB^D = B^D B = B^D A$. Also, $AB^D B^D = B^D B B^D = B^D$. Let m be any integer at least as large as the index of C so that $C^m = 0$. Then $A^m = B^m = B^D B^{m+1}$ and it follows that

$B^D = A^D$. Since B has index 0 or 1, $(B^D)^D = B$ and thus $B = C_A$. The conclusion follows.

Theorem 6. If A and A^s are spectral inverses, then $C_{A^s} = A^D$, and N_{A^s} is a spectral inverse of N_A such that

$$A^D N_{A^s} = N_{A^s} A^D = 0. \tag{30}$$

Proof. Since A and A^s are spectral inverses, they have the same generalized range space and the same 0-space. It follows that A^D, $(A^s)^D$, and C_{A^s} have the same 0-space. Since A^D and $(A^s)^D$ are reflexive generalized inverses of each other having the same generalized range and 0-spaces, the matrices AA^D and $A^s(A^s)^D$ are projections on the generalized range space along the 0-space. By uniqueness of the projection, $A^s(A^s)^D = AA^D$. Now, from Definition 5 for N_{A^s},

$$\begin{aligned}
A^D N_{A^s} &= A^D[I - A^s(A^s)^D]A^s \\
&= A^D A^s - A^D A^s(A^s)^D A^s \\
&= A^D A^s - A^D A A^D A^s \\
&= 0.
\end{aligned}$$

In a similar manner, $N_{A^s} A^D = 0$, thus (30) is established.

In view of $A = C_A + N_A$ and $A^D N_A = N_A A^D = 0$ [see problem 7(c)], appropriate multiplications yield

$$\begin{aligned}
A &= AA^s A = C_A C_{A^s} C_A + N_A N_{A^s} N_A = C_A + N_A, \\
A^s &= A^s A A^s = C_{A^s} C_A C_{A^s} + N_{A^s} N_A N_{A^s} = C_{A^s} + N_{A^s}.
\end{aligned}$$

Premultiplying these equations by AA^D and using $A^D N_A = N_A A^D = 0$ and (30) give

$$C_A C_{A^s} C_A = C_A, \quad C_{A^s} C_A C_{A^s} = C_{A^s}. \tag{31}$$

Hence also $N_A N_{A^s} N_A = N_A$, $N_{A^s} N_A N_{A^s} = N_{A^s}$, and thus C_{A^s} and N_{A^s} are reflexive generalized inverses of C_A and N_A, respectively. Since N_{A^s} is nilpotent, it is immediate that N_{A^s} is a spectral inverse of N_A (see problem 8).

Finally, it follows from (31) that C_{A^s} is a reflexive generalized inverse of C_A having the same row space and column space as C_A. From Theorem 4, A^D is also a reflexive generalized inverse with this property. Hence there exist matrices P and Q such that $C_{A^s} = A^D P$, $A^D = Q C_{A^s}$. Therefore

$$C_{A^s} = A^D P = A^D C_A A^D P = A^D C_A C_{A^s} = Q C_{A^s} C_A C_{A^s} = Q C_{A^s} = A^D,$$

the desired result. We now present the main result which is due to Greville [12].

Theorem 7. Two square matrices A and X are spectral inverses if and only if they satisfy the system of equations (17) to (20).

Proof. Assume A and X are spectral inverses with indices k_1 and k_2, respectively. Let $k \geq \max(k_1, k_2)$. Since $A = C_A + N_A$ with $C_A N_A = N_A C_A = 0$, we have $A^p = C_A{}^p + N_A{}^p$ $(p = 1, 2, \ldots)$. Since the index of N_A is k_1 and $C_{A^s} = A^D$, it follows that $A^k = C_A{}^k$, $X^k = (A^D)^k$. From Definition 4, and the fact that $AA^D = A^D A$, it is readily shown that A^k commutes with A^D and X^k commutes with C_A. Also, from $A^D N_A = N_A A^D = 0$, the definition of C_A, and (30), we have $A^k N_X = N_X A^k = 0$, $X^k N_A = N_A X^k = 0$. Thus writing A and X in the form $A = C_A + N_A$ and $X = C_X + N_X$, respectively, gives (19) and (20)

Conversely, assume A and X satisfy the system (17) to (20). Then, from (8) and (9) and the idempotency of AA^D,

$$XAA^D = X(AA^D)^{k+1} = XA^{k+1}(A^D)^{k+1} = A^k XA(A^D)^{k+1} = A^k(A^D)^{k+1} = A^D.$$

Let x be any λ-vector of A for some $\lambda \neq 0$. It can be shown that the column space of A^D is the generalized range space of both A and A^D. Thus x is in the column space of A^D, and therefore $x = A^D y$ for some y. Hence $XAx = XAA^D y = A^D y = x$. Therefore

$$X(A - \lambda I)x = (I - \lambda X)x. \tag{32}$$

In a similar manner, by interchanging the roles of A and X, we obtain

$$A(X - \mu I)z = (I - \mu A)z \tag{33}$$

for any μ-vector z of X for $\mu \neq 0$.

Repeated application of (32) and (33), using the fact that $AA^D = A^D A$, we obtain

$$X^p(A - \lambda I)^p x = (I - \lambda X)^p x \tag{34}$$

and

$$A^p(X - \mu I)^p z = (I - \mu A)^p z. \tag{35}$$

It follows from (34) that $(A - \lambda I)^p x = 0$ implies $(X - \lambda^{-1} I)^p x = 0$. Similarly, reasoning from (35) establishes the converse statement, and thus x is of the same grade as a λ-vector of A and as a λ^{-1} vector of X.

By applying similar reasoning to row vectors, one concludes that x is of the same grade as a row λ-vector of A and as a row λ^{-1} vector of X. It follows that X has the same row and column generalized range spaces as does A. Since the row generalized range space is the orthogonal complement of the column 0-space, A and X have the same generalized range space and the same 0-space and are therefore spectral inverses.

In view of (34) and (35), any spectral inverse more than satisfies property V.

3. STRONG SPECTRAL INVERSES

Since there is a large collection of spectral inverses for every square matrix of index greater than one, we impose further conditions to restrict the class of spectral inverses.

DEFINITION 6. Let A and A^s be spectral inverses. If in addition there is a non-singular matrix P such that the relations $A = P^{-1}JP$, $A^s = P^{-1}J^+P$ hold simultaneously, where J is a Jordan canonical form of A, then A^s is defined to be a strong spectral inverse of A.

We first establish that every square matrix A has a strong spectral inverse. Certainly, for any square matrix A, there is a nonsingular matrix P such that $A = P^{-1}JP$ where J is a Jordan canonical form of A. $X = P^{-1}J^+P$ is a reflexive generalized inverse of A and it follows from exercises 9 and 10 that X is a spectral inverse.

Although the class of strong spectral inverses is, in general, much more restricted than the class of spectral inverses, any matrix of index greater than one has an infinite set of strong spectral inverses. A unique member of the set is obtained once a particular P is specified such that $A = P^{-1}JP$, where J is a Jordan canonical form of A.

It certainly seems desirable to specify P so that as many properties of the regular inverse as possible are inherited by A^s. Since only partial results have been obtained, the reader is referred to the literature and encouraged toward the pursuit of knowledge in this area. See Appendix 2 for some notes and comments on spectral inverses.

EXERCISES

1. Prove that the following statements are equivalent.
   (a) The index of A is k.
   (b) k is the smallest integer for which $A^{k+1}A^D = A^k$.
   (c) k is the smallest integer such that A^k and A^{k+1} have the same rank.
   (d) A has a 0-vector of grade k but none of grade greater than k.
2. Let X be any reflexive generalized inverse of A. Then $XA = AX$ if and only if $X = A^D$.
3. Prove that $(A^D)^D = A^2A^D$. If A is singular, the index of A^D is 1.
4. (a) $(A^D)^D = A$ if and only if the index of A is 0 or 1.
   (b) For any square matrix A, $[(A^D)^D]^D = A^D$.
5. Prove that AA^D has the same rank as A^k where k is the index of A.
6. Prove that A and A^D are spectral inverses if and only if A is of index 0 or 1.
7. Prove that the nilpotent part N of any square matrix A of index k has the following properties.

(a) If k is 0 or 1, $N = 0$.

(b) For $k \geq 1$, $N^k = 0$.

(c) N satisfies the relations

$$A^D N = N A^D = 0,$$
$$C_A N = N C_A = 0,$$
$$(I - A A^D) N = N(I - A A^D) = N.$$

(d) rank A = rank C_A + rank N.

(e) Every 0-vector of A is a null vector of N, and every λ-vector of A for $\lambda \neq 0$ is a null vector of N.

(f) The index of N is k.

8. Prove that a spectral inverse of a nilpotent matrix A is any nilpotent matrix which is a reflexive generalized inverse of A.

9. Prove that for any Jordan matrix J, J^+ is a spectral inverse.

10. Prove that the spectral inverse relationship is preserved under a similarity transformation.

4

SPECIAL TOPICS

1. RANGE–HERMITIAN MATRICES

The concept of a range-hermitian matrix is a natural generalization of the concept of a normal matrix. Range-hermitian matrices were first studied by Schwerdtfeger and by Pearl [1–3], both of whom called them EPr matrices. The term range-hermitian was suggested by Greville.

DEFINITION 1. An $n \times n$ matrix A with elements from the complex field is range-hermitian if A and A^* have the same range space, that is, $R(A) = R(A^*)$.

We now establish several necessary and sufficient conditions for a matrix to be range-hermitian.

Theorem 1. Let A be an $n \times n$ matrix of rank r. Each of the following is necessary and sufficient for A to be range-hermitian:
  (a) $N(A) = N(A^*)$.
  (b) $R(A) = R(A^*)$.
  (c) A can be represented as

$$A = P\begin{pmatrix} D & DX^* \\ XD & XDX^* \end{pmatrix}P^*$$

where P is a permutation matrix and D is an $r \times r$ nonsingular matrix.
  (d) There exists a nonsingular matrix T such that

$$TAT^* = \begin{pmatrix} D & 0 \\ 0 & 0 \end{pmatrix}$$

where D is $r \times r$ and nonsingular.

31

(e) There is a nonsingular matrix N such that $A^* = NA$.

(f) There is a square matrix N such that $A^* = NA$.

Proof. We will show that (a) $\Rightarrow$ (b) $\Rightarrow$ (c) $\Rightarrow$ (d) $\Rightarrow$ (e) $\Rightarrow$ (f) $\Rightarrow$ (a).

Since $R(A^*) = N(A)^\perp$ and $R(A) = N(A^*)^\perp$, it follows that $N(A) = N(A^*)$ if and only if $R(A) = R(A^*)$. To establish (b) $\Rightarrow$ (c), note that there exists a permutation matrix Q such that

$$QAQ^* = B = \begin{pmatrix} D & E \\ F & G \end{pmatrix}$$

where D is an $r \times r$ nonsingular matrix. Since $[D, E]$ is of the same rank r as A and thus B, there is an $(n - r) \times r$ matrix X such that $[F, G] = X[D, E]$. Since A satisfies (a), it follows that

$$\begin{pmatrix} E \\ G \end{pmatrix} = \begin{pmatrix} D \\ F \end{pmatrix} X^*.$$

Let

$$R = \begin{pmatrix} I_r & 0 \\ -X & I_{n-r} \end{pmatrix}.$$

It follows that

$$\begin{pmatrix} D & 0 \\ 0 & 0 \end{pmatrix} = RBR^* = RQAQ^*R^*$$

so that

$$\begin{aligned} A &= Q^{-1}R^{-1}\begin{pmatrix} D & 0 \\ 0 & 0 \end{pmatrix}(Q^*R^*)^{-1} \\ &= Q^{-1}\begin{pmatrix} I_r & 0 \\ X & I \end{pmatrix}\begin{pmatrix} D & 0 \\ 0 & 0 \end{pmatrix}\begin{pmatrix} I_r & X^* \\ 0 & I \end{pmatrix}(Q^*)^{-1} \\ &= Q^*\begin{pmatrix} D & DX^* \\ XD & XDX^* \end{pmatrix}Q. \end{aligned}$$

Setting $P = Q^*$, (c) is established. Part (d) follows from (c) by setting

$$T = \begin{pmatrix} I_r & 0 \\ -X & I \end{pmatrix}P^*.$$

To establish (e), let

$$N = T^{-1}\begin{pmatrix} D^*D^{-1} & 0 \\ 0 & I \end{pmatrix}T.$$

Then

$$\begin{aligned} NA &= T^{-1}\begin{pmatrix} D^*D^{-1} & 0 \\ 0 & I \end{pmatrix}TT^{-1}\begin{pmatrix} D & 0 \\ 0 & 0 \end{pmatrix}(T^*)^{-1} \\ &= T^{-1}\begin{pmatrix} D^* & 0 \\ 0 & 0 \end{pmatrix}(T^*)^{-1} = A^*. \end{aligned}$$

Clearly, (e) implies (f). To show (f) implies (a), let $A^* = NA$. Let $X \in N(A)$. Then $A^*X = 0$ so that $N(A) \subseteq N(A^*)$. But since A and A^* have the same rank, $N(A) = N(A^*)$.

We now establish some relationships between range-hermitian matrices and generalized inverses.

Lemma 1. $A^* = A^*BA$ if and only if B is a generalized inverse of A and A is range-hermitian. Furthermore, $A^* = A^*BA$ if and only if $A^* = ABA^*$.

Proof. If B is a generalized inverse of A, then from the proof of Theorem 12 in Chapter 1, $N(BA) = N(A)$. If A is also range-hermitian, $N(BA) = N(A^*)$. Let X be an arbitrary vector in E^n. Then $X = Y + Z$ where $Y \in N(A)$ and $Z \in R(BA)$. Since $A^*BAX = A^*BAZ = A^*Z$ from the idempotency of BA, it follows that $A^*BA = A^*$. Conversely, if $A^*BA = A^*$, then $N(A) \subseteq N(A^*)$, hence $N(A) = N(A^*)$. Now $A^*BA = A^*$ implies that $A^*(I - BA) = 0$ and since $N(A) = N(A^*)$, this implies that $A(I - BA) = 0$ or $A = ABA$. In an analogous manner it can be shown that B being a generalized inverse of A and A being range-hermitian are necessary and sufficient for $A^* = ABA^*$.

Theorem 2. Let A be a range-hermitian matrix. There exist reflexive generalized inverses of A which are range-hermitian. In fact, $B = B_1 A^* B_1^*$, where B_1 is any reflexive generalized inverse of A, is such a matrix.

Proof. If $N(A) = N(A^*)$ and B_1 is a reflexive generalized inverse of A, then by Lemma 1, $A^*B_1A = A^*$ and $A^*B_1^*A = A$. Let $B = B_1 A^* B_1^*$, then $ABA = AB_1(A^*B_1^*A) = AB_1A = A$. That B and A have the same rank follows from the construction of B. Thus B is a reflexive generalized inverse of A. Clearly, $N(B_1^*) \subseteq N(B)$ and $N(B_1^*) \subseteq N(B^*)$. But B and B_1 have the same rank, therefore $N(B) = N(B^*)$.

2. PARTIAL ISOMETRICS

In this section several basic properties of unitary matrices are stated and generalizations for partial isometries are obtained. Let $\|x\|$ denote the Euclidean norm of a vector, that is, $\|x\|^2 = x^*x$.

DEFINITION 2. An $m \times n$ matrix A is called a partial isometry if the linear transformation $y = Ax$ for all $x \in N(A)^\perp$ preserves distances:

$$\|Ax_1 - Ax_2\| = \|x_1 - x_2\| \qquad \text{for all } x_1, x_2 \in N(A)^\perp. \tag{1}$$

If $N(A)$ consists only of the null vector, then $N(A)^\perp$ is E^n and A is unitary. Condition (1) can be readily shown to be equivalent to the following two conditions:

$$\|Ax\| = \|x\| \qquad \text{for all } x \in N(A)^\perp, \tag{2}$$

$$x_1^* A^* A x_2 = x_1^* x_2 \qquad \text{for all } x_1, x_2 \in N(A)^\perp. \tag{3}$$

A square matrix U is unitary if and only if $U^* = U^{-1}$. The generalization of this property for partial isometries is given in Theorem 3.

Lemma 2. An $m \times n$ matrix A is a partial isometry if and only if

$$A^*Ax = x \qquad \text{for all } x \in N(A)^\perp. \tag{4}$$

Proof. Let A be a partial isometry. From (2) we have $x^*A^*Ax = x^*x$ for all $x \in N(A)^\perp$. The general form of a vector A^*Ax satisfying this equation is $A^*Ax = x + y$ with $x^*y = 0$. Since y is uniquely expressible as $y = y_1 + y_2$ with $y_1 \in N(A)$, $y_2 \in N(A)^\perp$, we have $x^*y_1 = 0$. Also, $x^*y_2 = 0$ since $x^*y = 0$. Since $Ay_1 = 0$, $x^*A^*Ay = x^*A^*Ay_2$. But $x^*A^*Ay_2 = x^*y_2 = 0$ since A is a partial isometry and x, $y_2 \in N(A)^\perp$. It follows that $y^*A^*Ax = y^*(x + y) = y^*y = 0$, thus proving the necessity. Conversely, the isometric relation (2) follows immediately from (4).

Theorem 3. An $m \times n$ matrix A is a partial isometry if and only if $A^+ = A^*$.

Proof. If $A^+ = A^*$, then $A^+Ax = x + z$ where x is any vector in $N(A)^\perp$ and $z \in N(A)$. It follows that $x^*A^+Ax = x^*x = x^*A^*Ax$ for all $x \in N(A)^\perp$.

If A is a partial isometry, by Lemma 2, $A^*Ax = x$ for all $x \in N(A)^\perp$. Hence $AA^*Ax = Ax$ for all $x \in N(A)^\perp$. Expressing an arbitrary vector y in E^n as $y = x_1 + x_2$, with $x_1 \in N(A)$, $x_2 \in N(A)^\perp$, we obtain $AA^*Ay = AA^*Ax_2 = Ax_2 = A(x_1 + x_2) = Ay$. Hence $AA^*A = A$ and thus $A^*AA^* = A^*$. Since A^*A and AA^* are hermitian, $A^* = A^+$ follows from the uniqueness of A^+.

If U is unitary, so is U^*. It follows immediately from the fact that $(A^+)^+ = A$, and if A is a partial isometry, then so is A^*.

The eigenvalues of a unitary matrix have absolute value one. This is generalized to partial isometries in Theorem 4.

It follows from Theorem 3 that if A is a partial isometry, then all nonzero eigenvalues of A^*A are unity since $A^+A = A^*A$ is hermitian and idempotent. Furthermore, if A is square, its singular values, that is, the positive square roots of the eigenvalues of A^*A, are zero and one. Thus, from the above and the readily established inequality, we can conclude that the eigenvalues of a square partial isometry have absolute magnitudes on the closed interval $[0, 1]$, true for any square matrix:

$$\lambda_{\min}(A^*A) \leq |\lambda_i(A)|^2 \leq \lambda_{\max}(A^*A)$$

where $\lambda_i(B)$ denotes an eigenvalue of B.

Theorem 4. Let λ be an eigenvalue and x an associated eigenvector of a square partial isometry. Then one of the following cases holds:
  (1) $\lambda = 0$ if and only if $x \in N(A)$.
  (2) $|\lambda| = 1$ if and only if $x \in N(A)^\perp$.
  (3) $0 < |\lambda| < 1$ if and only if $x \notin N(A) \cup N(A)^\perp$.

Proof. With x, λ as given, the equation $Ax = \lambda x$ leads to $x^*A^*Ax = \lambda\bar{\lambda}x^*x$. If $\lambda = 0$, then $x^*A^*Ax = 0$ implies that $Ax = 0$ and hence $x \in N(A)$. If $x \in N(A)$, then since $x \neq 0$ it follows that $\lambda = 0$. In case (2), with $x \in N(A)^\perp$, $x^*A^*Ax = \lambda\bar{\lambda}x^*x$ compared to (2) leads to $|\lambda| = 1$, and conversely, if $|\lambda| = 1$, then $x^*A^*Ax = x^*x$ is the relation in (2), thus $x \in N(A)^\perp$. Case (3) is a consequence of the comments preceding this theorem by excluding cases (1) and (2).

If U is unitary, every matrix unitarily similar to U is unitary. This is generalized in Theorem 5.

Theorem 5. The matrix A is a partial isometry if and only if UAV is a partial isometry where U and V are unitary.

Proof. Let $P = UAV$ where U and V are any unitary matrices such that the product P is defined. Consider $PP^* = UAA^*U^*$. If A is a partial isometry, $AA^* = AA^+$ is idempotent hence PP^* is idempotent. This implies P is a partial isometry. See problem 6. Similarly, if P is a partial isometry, then $A = U^*PV^*$ is a partial isometry.

In particular, if A is a partial isometry, then every matrix unitarily similar to A is a partial isometry.

An $n \times n$ matrix is unitary if and only if its columns (rows) form an orthonormal basis of E^n. Certainly, neither the rows nor the columns of a singular $n \times n$ matrix can form an n-dimensional orthonormal basis, however, a generalization is given in the next theorem.

Theorem 6. An $n \times n$ partial isometry A of rank r is unitarily similar to a matrix whose non-null columns (or rows) determine an r-dimensional orthonormal basis of a subspace of E^n. Conversely, any square matrix unitarily similar to a matrix whose non-null columns (or rows) form a set of r orthonormal vectors is a partial isometry of rank r.

Proof. There exists a unitary matrix U such that

$$U^*(A^*A)U = \begin{pmatrix} I_r & 0 \\ 0 & 0 \end{pmatrix}.$$

Letting $B = U^*AU$ we have

$$B^*B = \begin{pmatrix} I_r & 0 \\ 0 & 0 \end{pmatrix}.$$

Partitioning B into its columns, for example, $B = [B_1, B_2, \ldots, B_n]$, and forming the product B^*B, we have $B_i^*B_j = \delta_{ij}$, $i, j = 1, 2, \ldots, r$, where

$$\delta_{ij} = \begin{cases} 1 & \text{if } i = j, \\ 0 & \text{if } i \neq j, \end{cases}$$

and $B_p^* B_q = 0$, $p, q = r + 1, \ldots, n$ since

$$B^*B = \begin{pmatrix} I_r & 0 \\ 0 & 0 \end{pmatrix}.$$

Conversely, $B_{r+1} = B_{r+2} = \cdots = B_n = 0$, hence any vector $x \in N(B)^\perp$ has its last $n - r$ components zero, thus $B^*Bx = x$ for all $x \in N(B)^\perp$. Hence B is a partial isometry and by Theorem 5 A is a partial isometry.

It is immediate from the proof of Theorem 6 that any $m \times n$ or $n \times n$ matrix whose nonzero columns (or rows) form an orthonormal set of r vectors is a partial isometry of rank r.

The next property of unitary matrices to be considered is that every normal matrix A is unitarily similar to a diagonal matrix. This is generalized in the following theorem.

Theorem 7. Let B be an $n \times n$ normal matrix of rank r, then there exists an $n \times r$ partial isometry A of rank r such that $A^*BA = D = \text{diag}(\lambda_1, \ldots, \lambda_r)$ where the λ_i are the nonzero eigenvalues of B.

Proof. Let U be a unitary matrix that diagonalizes B. If $r = n$, we are done. If $r < n$, let

$$U^*BU = \begin{pmatrix} D & 0 \\ 0 & 0 \end{pmatrix}.$$

Pre- and postmultiplying by

$$(I_r, 0) \quad \text{and} \quad \begin{pmatrix} I_r \\ 0 \end{pmatrix},$$

respectively, and letting

$$A = U \begin{pmatrix} I_r \\ 0 \end{pmatrix},$$

give the desired conclusion.

It is readily seen that Theorem 7 could be generalized so that $D = \text{diag}(\lambda_1, \ldots, \lambda_k)$ where $\{\lambda_1, \ldots, \lambda_k\}$ is any subset of the set of eigenvalues of B.

A product of unitary matrices is always unitary. That this is not always true for partial isometries follows from problem 7. However, necessary and sufficient conditions for the product of two partial isometries to be a partial isometry are established in Theorem 8.

Theorem 8. Let the $m \times n$ matrix A and the $n \times p$ matrix B be partial isometries with $P = AB$. Then P is a partial isometry if and only if the image of $N(P)^\perp$ under B is contained in $N(A)^\perp$.

Proof. Let $x \neq 0$ be a vector in $N(P)^{\perp} \subseteq N(B)^{\perp}$, and consider the equation $x^*P^*Px = x^*B^*A^*ABx$. If $R(B) \subseteq N(A)^{\perp}$, then $Bx \in N(A)^{\perp}$, hence from Lemma 2, $A^*ABx = Bx$. Thus $x^*P^*Px = x^*B^*Bx = x^*x$ and sufficiency is established.

Assume now that $x^*P^*Px = x^*x$ for $x \in N(P)^{\perp}$ but the image of $N(P)^{\perp}$ under B is not contained in $N(A)^{\perp}$. Let $y = Bx$ with $y \notin N(A)^{\perp}$, then isometry with respect to A no longer holds, that is, $y^*A^*Ay \neq y^*y$. Thus $x^*P^*Px \neq x^*B^*Bx = x^*x$ which completes the proof.

If A is a nonsingular complex matrix, there exist unique matrices H and U such that H is positive definite, U is unitary, and $A = HU$. This is known as the polar decomposition of A. This result is generalized in the sequel, where it is understood that an $m \times n$ matrix is diagonal if its elements a_{ij} are all zero unless $i = j$.

Lemma 3. Every $m \times n$ matrix A can be factored as $A = UDV$ where D is a diagonal matrix in the sense discussed above with real, non-negative elements, and U, V are unitary.

Proof. Since AA^* is a hermitian non-negative definite matrix, there is a set of m orthonormal vectors such that

$$AA^*x_i = \lambda_i^2 x_i; \quad i = 1, 2, \ldots, m. \tag{5}$$

The numbers λ_i^2 are the eigenvalues of AA^*. The numbers λ_i are thus real and may be taken to be non-negative.

In a similar manner, there is a set of n orthonormal vectors such that

$$A^*Ay_i = \sigma_i^2 y_i; \quad i = 1, 2, \ldots, n. \tag{6}$$

Arrange the numbering of these vectors so that

$$\lambda_1 \geq \lambda_2 \geq \cdots \geq \lambda_r > 0, \quad \lambda_{r+1} = \cdots = \lambda_m = 0,$$

$$\sigma_1 \geq \sigma_2 \geq \cdots \geq \sigma_s > 0, \quad \sigma_{s+1} = \cdots = \sigma_n = 0.$$

Certainly, $r = s$ and $\lambda_i = \sigma_i$ whenever $i \leq s$. Moreover, the s vectors defined by $y_i = \lambda_i^{-1}A^*x_i$, $i \leq s$, will satisfy (6) with $\sigma_i = \lambda_i$. The vectors $x_1, \ldots, x_m$ can be taken as the columns of a unitary matrix U. Let $y_1, \ldots, y_s$ be the first s columns of an $n \times n$ matrix V^*, and fill in the remaining columns to make V^* unitary. These matrices U and V, with D taken to be the matrix having $\lambda_1, \ldots, \lambda_r$ in the first r diagonal positions and zeroes elsewhere, then satisfy the requirements of the lemma.

Combining the results of Theorems 5 and 6, we find that if D is an $m \times n$ matrix, with ones in the first r diagonal positions and zero entries elsewhere, then an $m \times n$ matrix A of rank r is a partial isometry if and only if $A = UDV$ where U and V are unitary. It is now immediate that a square matrix A is a partial isometry if and only if $A = QE$ and if and only if $A = FQ$, where Q is unitary and E and F are orthogonal projections since we can write $A = (UV)(V^*DV) = QE = (UDU^*)UV = FQ$, where $Q = UV$, $E = V^*DU$, and $F = UDU^*$.

Theorem 9. Every matrix of rank r can be factored as $A = P_1DV = UDP_2 = P_1DP_2$ where U, V are unitary, P_1 and P_2 are partial isometries, and D has positive entries on the first r diagonal positions and zero entries elsewhere.

Proof. Let A be factored as $A = UDV$ in Lemma 3. Thus we have $A = (UDD^+)DV = UD(D^+DV) = (UDD^+)D(D^+DV)$. Letting $P_1 = UDD^+$, $P_2 = D^+DV$, it follows that P_1, P_2 are partial isometries since DD^+ and D^+D are orthogonal projections and U, V are unitary.

Corollary 1. Any square matrix A can be factored as $A = PH$ where P is a partial isometry and H is positive semidefinite and hermitian.

Proof. From Theorem 9, if $A = P_1DV$ is square, we have $A = P_1V(V^*DV)$ where $H = V^*DV$ is positive semidefinite and $P = P_1V$ is a partial isometry.

We will now establish conditions that uniquely determine P and H in the above corollary.

Theorem 10. Let the $m \times n$ matrix A be factored as $A = PH$ where P is a partial isometry and H is positive semidefinite, hermitian.

    (a) If $P^*A = H$, H is uniquely determined.

    (b) If $N(P) = N(H)$, then both P and H are uniquely determined.

Proof. If $P^*A = H$, then $A^*A = P^*AP^*PP^*A = P^*AP^*A = H^2$. Thus H is the unique positive semidefinite square root of A^*A.

To establish (b), let $A = P_1H_1 = P_2H_2$ be any two factorizations of A and assume $N(P_i) = N(H_i)$, $i = 1, 2$. It follows that $R(P_i^*) = R(H_i)$. Thus $P_i^*A = H_i$. From (a) we have $H_1 = H_2 = H$, and thus $P_1H = P_2H$. This implies $P_1x = P_2x$ when $x \in R(H)$. Since $N(P_1) = N(P_2) = N(H)$, we have $P_1y = P_2y$ for $y \in N(H)$. Hence $P_1 = P_2$. Moreover, it should be noted that $P = AH^+$.

3. DIFFERENTIABLE GENERALIZED INVERSES

If a function is such that its first k derivatives exist and are continuous, we write $f \in C^k$. If the entries a_{ij} of A are such that $a_{ij} \in C^k$ for all i, j, we write $A \in C^k$.

We first establish the existence of differentiable generalized inverses.

Lemma 4. A continuous projection $A(t)$ defined on an interval has constant rank.

Proof. Let $r(t)$ be the rank of $A(t)$. Clearly, the determinant $\det(A + I) = 2^{r(t)}$. Since $\det(A + I)$ is continuous, it follows that $r(t)$ is constant since a continuous integral-valued function on an interval is constant.

From Theorem 1 in Chapter 3 we find that for a given matrix A there is a unique reflexive generalized inverse which commutes with A if and only if $r(A) = r(A^2)$.

The next lemma concerns the differentiability of this reflexive generalized inverse.

Lemma 5. Let $A \in C^k$ on an interval $[a, b]$ such that $r[A(t)] = r[A^2(t)] = r$ for each $t \in [a, b]$. If $B(t)$, for each $t \in [a, b]$ is the unique reflexive generalized inverse of $A(t)$ which commutes with $A(t)$, then $B \in C^k$ on $[a, b]$.

Proof. With A as specified, the Jordan canonical form of $A(t)$ is

$$\begin{pmatrix} A_1(t) & 0 \\ 0 & 0 \end{pmatrix}$$

where $A_1(t)$ is an r-square nonsingular matrix. The characteristic polynomial of $A(t)$ has coefficients in C^k and is of the form $\lambda^{n-r} f(\lambda)$, where $f(\lambda)$ is the characteristic polynomial of $A_1(t)$. From $f(\lambda)$ we can construct a polynomial $p(\lambda)$, with coefficients in C^k, such that $A_1^{-1}(t) = p[A_1(t)]$. The matrix $B_1(t) = p[A(t)]$ is a generalized inverse of $A(t)$. The matrix $B = B_1 A B_1$ is a reflexive generalized inverse of A, commutes with A, and clearly $B \in C^k$.

We now establish the existence of differentiable generalized inverses.

Theorem 11. Let $A \in C^k$ on $[a, b]$. If $r(A)$ is constant on $[a, b]$, then A^+ is in C^k on $[a, b]$. If $r(A)$ is not constant on $[a, b]$, then no generalized inverse of A can be continuous on $[a, b]$.

Proof. If $r(A)$ is constant on $[a, b]$, the matrix $H = A^*A$ satisfies the hypothesis of Lemma 5. Thus there is a matrix $W(t)$ which is the unique reflexive generalized inverse of $H(t)$ that commutes with $H(t)$, and $W \in C^k$ on $[a, b]$ By considering the defining equations for W and their conjugate transposes it follows that W is hermitian. Thus $W = H^+$. Hence $B = WA^* = A^+$ and clearly $B \in C^k$.

Now let B be any generalized inverse of A. If B is continuous on $[a, b]$, then, by Lemma 4, the continuous projection AB has constant rank on $[a, b]$. However, $r(AB) = r(A)$, and A has constant rank.

It is well known that if A is a nonsingular matrix having entries that are differentiable functions of t, then the derivative of A and that of A^{-1} are related by

$$\frac{dA}{dt} = -A \frac{dA^{-1}}{dt} A, \tag{7}$$

$$\frac{dA^{-1}}{dt} = -A^{-1} \frac{dA}{dt} A^{-1}. \tag{8}$$

We now investigate the relations analogous to (7) and (8) that hold for a singular differentiable matrix A and a differentiable generalized inverse A^g of A.

Lemma 6. If P is a differentiable projection, then $P(dP/dt)P = 0$.

Proof. Since P is a projection, $P = P^2$, and we have

$$\frac{dP}{dt} = \frac{dP}{dt} P + P \frac{dP}{dt}.$$

Left multiplication by P, using the fact that $P = P^2$, yields the desired result.

The next theorem furnishes extensions of (7) and (8) to a singular matrix.

Theorem 12. If $A \in C^1$ and $A^g \in C^1$, then

$$AA^g \frac{dA}{dt} A^g A = -A \frac{dA^g}{dt} A \tag{9}$$

and if $A^r \in C^1$, then

$$A^r A \frac{dA^r}{dt} AA^r = -A^r \frac{dA}{dt} A^r. \tag{10}$$

Proof. Let $A^g \in C^1$. By Lemma 6, with $P = AA^g$, we have $P(dP/dt)A = P(dP/dt)PA = 0$. Thus $P(dP/dt)A = P[(dA/dt)A^g + A(dA^g/dt)]A = P(dA/dt)A^g A + A(dA^g/dt)A = 0$, from which (9) follows.

Let $A^r \in C^1$. Pre- and postmultiplying (9) by A^r, replacing A^g by A^r, yield $A^r AA^r(dA/dt)A^r AA^r = -A^r A(dA^r/dt)AA^r$ which reduces to (10).

It should be noted from the above proof that (10) follows from (9) only if A is a generalized inverse of A^g, and A^g is a generalized inverse of A.

We will now establish conditions under which a differentiable generalized inverse of A can be substituted for A^{-1} in (7), and conditions under which a differentiable reflexive generalized inverse of A can be substituted for A^{-1} in (8).

Theorem 13. (a) If $A \in C^1$ and $A^g \in C^1$, then $dA/dt = -A(dA^g/dt)A$ if and only if $[d(AA^g)/dt]A = A[d(A^g A)/dt] = 0$.

(b) If $A^r \in C^1$, then $dA^r/dt = -A^r(dA/dt)A^r$ if and only if $[d(A^r A)/dt]A^r = A^r[d(AA^r)/dt] = 0$.

Proof. If $[d(AA^g)/dt]A = A[d(AA^g)/dt] = 0$, then from $A = AA^g A$ we have $dA/dt = AA^g(dA/dt) = (dA/dt)A^g A$. But then (9) reads $dA/dt = -A(dA^g/dt)A$. Conversely, if $dA/dt = -A(dA^g/dt)A$, left multiplication by AA^g and right multiplication by $A^g A$ yield, respectively, that $AA^g(dA/dt) = dA/dt$ and $(dA/dt)A^g A = dA/dt$. The result now follows from $dA/dt = [d(AA^g)/dt]A + AA^g(dA/dt) = A[d(A^g A)/dt] + (dA/dt)A^g A$. The second part of the theorem is established in an analogous manner and is left as an exercise.

EXERCISES

1. Let A be a square matrix of order n and rank r. Prove that each of the following is necessary and sufficient for A to be range-hermitian, thus extending the conclusion of Theorem 1.

   (a) There exists a unitary matrix U such that $UAU^* = \begin{pmatrix} D & 0 \\ 0 & 0 \end{pmatrix}$ where D is $r \times r$ and nonsingular.

   (b) A is the matrix of a linear transformation T on E^n for which there are mutually orthogonal subspaces V_1 and V_2 of E^n such that $T(V_1) = V_1$, $T(V_2) = 0$ with V_1 having dimension r, and $E^n = V_1 \oplus V_2$.

   (c) Let A_i and A^i be the ith row and ith column, respectively, of A. Then $\sum_{i=1}^{n} \alpha_i A_i = 0$ if and only if $\sum_{i=1}^{n} \bar{\alpha}_i A^i = 0$ (α_i, $i = 1, \ldots, n$ are scalars).

2. Prove that if A is range-hermitian, then $A^n (n = 1, 2, \ldots)$ is range-hermitian.

3. Show that every square matrix is a product of three range-hermitian matrices.

4. Prove that a normal matrix is range-hermitian.

5. Let $A = \begin{pmatrix} 1 & 1 \\ 1 & 1 \end{pmatrix}$, $B = \begin{pmatrix} 0 & 0 \\ 0 & 1 \end{pmatrix}$. Show that A and B are range-hermitian but AB is not.

6. Prove that each of the following four conditions is necessary and sufficient for a linear transformation A to be a partial isometry.

   (a) $AA^*A = A$.

   (b) $A^*AA^* = A^*$.

   (c) AA^* is a projection.

   (d) A^*A is a projection.

7. Show that A and B are partial isometries but AB and BA are not, where

$$A = \begin{pmatrix} 1 & 0 & 0 \\ 0 & \sqrt{3}/2 & 0 \\ 0 & 1/2 & 0 \end{pmatrix} \quad \text{and} \quad B = \begin{pmatrix} 1 & 0 & 0 \\ 0 & \sqrt{2}/2 & 0 \\ 0 & \sqrt{2}/2 & 0 \end{pmatrix}.$$

8. Let $A = PH$ where P is a partial isometry and H is hermitian. Prove that the condition $N(P) = N(H)$ is equivalent to the conditions $P^*A = H$ and $r(P) = r(H)$.

9. Establish the relations between the derivatives of A and A^{-1} given in (7) and (8). Show that (8) can always be obtained from (7) and conversely.

10. Establish part (b) of Theorem 13.

11. Prove that if A and A' are differentiable and commute, if either of the conditions $\dfrac{dA}{dt} = -A \dfrac{dA^r}{dt} A$ or $\dfrac{dA^r}{dt} = -A \dfrac{dA}{dt} A^r$ is satisfied, so is the other.

12. Show that (7) and (8) follow as special cases of Theorem 13.

13. Show that $A = \begin{pmatrix} \cos t & \sin t \\ 0 & 0 \end{pmatrix}$ has no continuous, real, reflexive generalized inverse which is range-hermitian on the interval $[0, \pi]$. *Hint:* Assume $N(A') = N(A'^*)$ and show that A' is not continuous.

5

SOLVING SYSTEMS OF LINEAR EQUATIONS

1. INTRODUCTION

Early in Chapter 1 we noted in Lemma 1 that a necessary and sufficient condition for the matrix equation $AXB = C$ to have a solution is $AA^+CB^+B = C$, in which case the general solution is $X = A^+CB^+ + Z - A^+AZBB^+$ where Z is arbitrary.

If A is an $m \times n$ matrix, the notation $\|A\|$ denotes the non-negative square root of the sum of squares of the moduli of the elements of A. Note that $\|A\|^2 = \operatorname{tr} A^*A$ and $\|A\| > 0$ unless $A = 0$, then $\|A\| = 0$.

DEFINITION 1. The matrix X_0 is a best approximate solution of the equation $f(X) = G$ if for all X, either

$\quad$ (i) $\quad \|f(X) - G\| > \|f(X_0) - G\|$ $\quad$ or

$\quad$ (ii) $\quad \|f(X) - G\| = \|f(X_0) - G\|$ $\quad$ and $\quad \|X\| \geq \|X_0\|$.

Theorem 1. The best approximate solution of the equation $AX = B$ is $X_0 = A^+B$.

Proof. It is readily established for matrices P and Q that

$$\|AP + (I - AA^+)Q\|^2 = \|AP\|^2 + \|(I - AA^+)Q\|^2. \tag{1}$$

In particular, then

$$
\begin{aligned}
\|AX - B\|^2 &= \|A(X - A^+B) + (I - AA^+)(-B)\|^2 \\
&= \|A(X - A^+B)\|^2 + \|(I - AA^+)(-B)\|^2 \\
&= \|AX - AA^+B\|^2 + \|AA^+B - B\|^2 \\
&\geq \|AA^+B - B\|^2.
\end{aligned}
$$

Equality holds only when $\|AX - AA^+B\| = 0$ or

$$AX = AA^+B. \tag{2}$$

If one replaces A by A^+ in (1) and uses the fact that $AA^+A = A$, one can deduce $\|A^+B + (I - A^+A)X\|^2 = \|A^+B\|^2 + \|(I - A^+A)X\|^2$. Now if (2) holds, A^+AA^+ gives $\|X\|^2 = \|A^+B\|^2 + \|X - A^+B\|^2$, which is minimal if $\|X - A^+B\| = 0$, or $X - A^+B = 0$ implying $X_0 = A^+B$.

Corollary 1. The best approximate solution of $AX = I$ is $X = A^+$.

Corollary 2. The best approximate solution x_0 of

$$Ax = y, \tag{3}$$

where x and y are vectors, is the vector $x_0 = A^+y$.

It is important to note that x_0 in Corollary 2 is simply the so-called least squares solution of (3), which is equivalent to that solution which minimizes the l_2 norm.

2. THE CONCEPT OF A *p-q* GENERALIZED INVERSE

In Section 1 of this chapter we defined what we called a best approximate solution of a set of linear equations making use of the l_2 norm. A natural extension of this definition incorporating strictly convex norms (i.e., norms whose unit spheres contain no line segments on their surface) is needed and is basic to the discussion that follows.

DEFINITION 2. Let $(V, \|\cdot\|_q)$, $(U, \|\cdot\|_p)$ be finite dimensional normed linear spaces with strictly convex norms p and q, and A be a linear transformation from V into U. For $y \in U$, a best approximate solution of $Ax = y$ is any x_0 in V such that for every other x in V either

    (i) $\|Ax_0 - y\|_p < \|Ax - y\|_p$ or

    (ii) $\|Ax_0 - y\|_p = \|Ax - y\|_p$ and $\|x_0\|_q < \|x\|_q$.

The existence of the best approximate solution with respect to the p and q norms is assured by the following theorem.

Theorem 2. For each y in U and every pair of strictly convex norms there exists a unique best approximate solution of $Ax = y$.

Proof. Since p is a strictly convex norm, there exists a unique $y_0 \in R(A)$ such that $\|y_0 - y\|_p \leq \|Ax - y\|_p$ for any $x \in V$. If z is any best approximate solution, clearly we must have $Ax = y_0$. The solutions to $Ax = y_0$ are precisely the elements of $z_0 + N(A)$ where $Az_0 = y_0$. By strict convexity there is a unique $x_0 \in z_0 + N(A)$ such that $\|x_0\|_q < \|x\|_q$ for all $x \in z_0 + N(A)$, $x \neq x_0$. The proof of this theorem follows directly from the fact that the norms are strictly convex. Since l_1 and l_∞ norms are not strictly convex, the uniqueness may be lost. The l_p norms $1 < p < \infty$ form a class of strictly convex norms.

The following definition and properties of a projection mapping, with respect to a strictly convex norm, are helpful in formulating a definition of a p-q generalized inverse.

DEFINITION 3. Let M be a subspace of U. For each $x \in U$, $E_M(x)$ is that element y of M which minimizes the distance between x and the set M with respect to the p-norm. That is, $\|x - y\|_p$ is minimum for $y \in M$, when $y = E_M(x)$. The mapping E_M is called the p-projection of U onto M. The following properties of the mapping E are easily established from the definition (it should be noted that $E_M = E$ may fail to be linear).
  (a) $E(\lambda x) = \lambda E(x)$ for any scalar λ.
  (b) $E^2 = E$.
  (c) $Ex = x$ if and only if $x \in M$.
  (d) $E(x + Y) = E(x) + y$ for $x \in U$, $y \in M$.
  (e) $E[x + E(Y)] = E(x) + E(y)$ for all $x, y \in U$.
  (f) $E[x - E(x)] = 0$ for all $x \in U$.

The definition of the p-q generalized inverse follows naturally from Theorem 3.

Theorem 3. Let A be a linear transformation mapping V into U. Let E be the p-projection on $R(A)$, and F be the q-projection on $N(A)$. The best approximate solution x_0, in the sense of Definition 2, is given by $x_0 = B(Y)$ where

$$B = (I - F)A^g E, \tag{4}$$

and A^g is any generalized inverse of A.

Proof. For each $y \in U$, $E(y)$ is the closest point in $R(A)$ to y with respect to the p-norm. Since $E(y)$ is in $R(A)$, $A^g E(y)$ satisfies the equation $Ax = E(y)$ where A^g is any linear mapping such that $AA^g A = A$. Since $AF = 0$, note that $B(y)$ also satisfies $Ax = E(y)$ since $AB(y) = A[(I - F)A^g E](y) = AA^g E(y)$. Thus $A^g E(y) - B(y) \in N(A)$. Now $FA^g E(y)$ is the closest point in $N(A)$ to $A^g E(y)$ so that $(I - F)A^g E(y)$ is the closest point in $B(y) + N(A)$ to zero. But $B(y)$ is unique with respect to the properties: (i) $AB(y)$ is the closest point in $R(A)$ to y, and (ii) $B(y)$ is the closest point to zero satisfying (i). Thus $B(y) = (I - F)A^g E(y)$ for every y. Thus the theorem is proved.

DEFINITION 4. The mapping B (not necessarily linear) as defined in (4) is said to be the p-q generalized inverse of the linear mapping A.

One can obtain the following useful properties directly by applying the definition of E and F:

$$AF = 0, \tag{5}$$
$$EA = A, \tag{6}$$
$$BE = A, \tag{7}$$
$$AA^g E = E, \tag{8}$$
$$AB = E, \tag{9}$$
$$BA = (I - F)A^g A, \tag{10}$$
$$ABA = A, \tag{11}$$
$$BAB = B. \tag{12}$$

Corollary 3. Let $q = 2$, then $F = I - A^+ A$ and $B = A^+ E$.

Corollary 4. Let $p = 2$, then $E = AA^+$ and $B = (I - F)A^+$.

Corollary 5. Let $p = q = 2$, then $B = A^+$.

Corollary 6. Let the rank of A be $n \le m$ where A is $m \times n$, then $F = 0$ and $B = A^g E$ for all q.

It is important to note that if A^p denotes the *p-q* generalized inverse of A, the symbol $(A^p)^p$ has no meaning in the context here since $A^p = B$ is not a linear map except when $p = q = 2$. A question that naturally arises is: what are the restrictions on A or the norms, p, q so that B is a linear mapping?

In case we restrict our attention to the l_p norms, it is not difficult to see that E_M is linear for all subspaces M, only in the case $p = 2$. This is shown by the following example in three dimensions.

In the vector space V_3 of ordered triples, let M be the subspace spanned by the vector $(1, 1, 1)$. Suppose $E = E_M$ is linear, where the norm is the l_p norm, $1 < p < \infty$. By symmetry we see that $E(1, 0, 0) = E(0, 1, 0) = E(0, 0, 1)$. Hence $E(3, 0, 0) = 3E(1, 0, 0) = E(1, 1, 1) = (1, 1, 1)$ since $(1, 1, 1) \in M$. That is, $\|(\lambda - 3, \lambda, \lambda)\|$ is minimized uniquely by $\lambda = 1$. We see that $f(\lambda) = \|(\lambda - 3, \lambda, \lambda)\|^p$ has a unique minimum at $\lambda = 1$, and indeed this is the only point for which $df(\lambda)/d\lambda = 0$. Note that for $p > 1$, $d|\lambda|^p/d\lambda = p\lambda|\lambda|^{p-2}$ (even at $\lambda = 0$). It follows that $f'(\lambda) = p(\lambda - 3)|\lambda - 3|^{p-2} + 2p|\lambda|^{p-2}$. Thus we have $0 = f'(1) = -2p2^{p-2} + 2p$ from which it follows that $2^{p-2} = 1$. That is, $p = 2$.

In view of the known facts about the l_2 norm we can state the following result.

Theorem 4. Consider the space V_n with the l_p norm, $1 < p < \infty$. E_M is linear for every subspace M of V_n if and only if $n \le 2$ or $p = 2$.

Corollary 7. For $m \ge 3$, $n \ge 3$, consider the spaces V_n and V_m with the l_q and l_p norms, respectively. B is linear for every $m \times n$ matrix A if and only if $p = q = 2$.

Lemma 1. Let M be a hyperplane contained in the normed linear vector space U, then E is a linear transformation.

Proof. Since M is a hyperplane, its dimension is $n - 1$ where the dimension of U is n. Let y_1 and y_2 be distinct vectors belonging to U, then y_1 and y_2 can be written as $y_1 = \lambda_1 y + m_1$ and $y_2 = \lambda_2 y + m_2$ where λ_1 and λ_2 are scalars. y is any vector not in M, and m_1 and m_2 are vectors belonging to M. Consider $E(y_1 + y_2) = E(\lambda_1 y + m_1 + \lambda_2 y + m_2) = E[(\lambda_1 + \lambda_2)y + m_1 + m_2]$. Properties (a), (d), and (e) of the transformation imply that $E(y_1 + y_2) = (\lambda_1 + \lambda_2)E(y) + m_1 + m_2 = [\lambda_1 E(y) + m_1] + [\lambda_2 E(y) + m_2] = E(y_1) + E(y_2)$, the desired result.

Theorem 5. If A is an $(n + 1) \times n$ matrix of rank n, then B is a linear transformation for all p.

Proof. Since $N(A) = \{0\}$, that is, $F = 0$, the matrix A imposes the condition of Lemma 1, hence E is a linear transformation. It follows easily that B is a linear transformation.

The operator E is the transformation that defines the vector x such that $Q = [\sum_{i=1}^{n} |y_i - A_i x|^p]^{1/p}$ is minimal where A_i is the ith row of the $(m \times n)$ matrix A.

Consider the set of equations

$$\frac{\partial Q}{\partial x_j} = \frac{1}{p} [Q^p]^{1/p - 1} \frac{\partial Q^p}{\partial x_j} = 0.$$

It is clear that

$$[Q^p]^{1/p - 1} = \left[\sum_{i=1}^{n} |y_i - A_i x|^p \right]^{1/p - 1} \geq 0,$$

and $\partial Q/\partial x_j$ may fail to exist if $|y_i - A_i x| = 0$ for all $i = 1, 2, \ldots, n$, that is, $y \in R(A)$. But in this case there exists an x_0 such that $Q \equiv 0$, hence x_0 is the desired value of x which minimizes Q. It follows that if $y \notin R(A), [Q^p]^{1/p - 1} > 0$ and Q is minimized if and only if Q^p is minimized.

Let $Q^p = \sum_{i=1}^{n} |y_i - A_i x|^p$ and

$$f_j = \frac{\partial Q^p}{\partial x_j} = - \sum_{i=1}^{n} p(y_i - A_i x)|y_i - A_i x|^{p - 2} (a_{ij}).$$

Also, the matrix of the second partial derivatives of Q^p is

$$D = \left\{ \frac{\partial f_j}{\partial x_j'} \right\} = p(p - 1)A^T W A$$

where W is a diagonal matrix defined by $W = \text{diag}\{|y_1 - A_1 x|^{p - 2}, \ldots, |y_n - A_n x|^{p - 2}\}$. In many practical cases $r(A) = n \leq m$, and the matrix D is positive definite since the diagonal elements of W are all positive. This statement can be supported by the heuristics in the following paragraph, which may be omitted by the uninitiated in probability theory.

Let $Y_A = \{y_i \text{ such that } y_i - A_i x = 0 \text{ for at least one } i = 1, 2, \ldots, n\}$ and $G(y)$ be the frequency distribution function of Y defined on a region of non-zero measure in Y. Since $m > n$, Y_A is the finite union of linear manifolds whose dimensions are less than n. Hence $\int_{Y_A} dG(y) = 0$, that is, $\Pr[y \in y_A] = 0$. In practice then W is almost surely positive definite.

Consider the equation $f_i(u_1, \ldots, u_m; v_1, \ldots, v_n) = 0$ $(i = 1, \ldots, n)$. When convenient, the sets $(u_1, \ldots, u_m; v_1, \ldots, v_n)$ and $(u_1, \ldots, u_m)$, $(v_1, \ldots, v_n)$ are denoted more simply by (u, v) and u, v, respectively; and the notations (a, b) and a, b in the following theorem will indicate points of similar types. By $D(u, v)$ is meant the functional determinant

$$D(u, v) = \frac{\partial(f_1, \ldots, f_n)}{\partial(v_1, \ldots, v_n)} = \left| \frac{\partial f_i}{\partial v_k} \right|.$$

The neighborhood $(a, b)_\varepsilon$ is the totality of points (u, v) satisfying the inequalities $|u_r - a_r| \le \varepsilon$, $|v_i - b_i| \le \varepsilon (r = 1, \ldots, m; \ i = 1, \ldots, n)$, and $a_\varepsilon, b_\varepsilon$ have similar meanings. With these understandings as to notations the theorem we need follows:

The Implicit Function Theorem (Bliss, pp. 269–270). Let $p = (a, b)$ be a point with the following properties:

(1) The functions $f_i(u, v)$ are continuous and have continuous derivatives $\partial f_i / \partial v_k$ in a neighborhood N of p.

(2) $f_i(a, b) = 0$ $(i = 1, \ldots, n)$.

(3) $D(a, b) \ne 0$.

Then there exists a set of functions $v_i(u_1, \ldots, u_m)$ single valued and continuous in a neighborhood a_δ, having the following properties:

(1) The points $[u, v(u_1, \ldots, u_m)]$ which they define are in N and satisfy the equation $f_i = 0$ $(i = 1, 2, \ldots, n)$.

(2) There exists a constant ε such that for each u in a_δ the set $[u, v(u_1, \ldots, u_m)]$ is the only solution (u, v) of the equation $f_i = 0$ satisfying the inequalities $v_i(u_1, \ldots, u_m) - \varepsilon < v_i < v_i(u_1, \ldots, u_m) + \varepsilon$.

(3) $v_i(a_1, \ldots, a_m) = b_i$ $(i = 1, \ldots, n)$.

(4) In a sufficiently small neighborhood a_δ the functions $v_i(u_1, \ldots, u_m)$ have continuous partial derivatives of as many orders as are possessed by functions f_i in the neighborhood N.

Let $f_i = \partial Q^p / \partial x_i$, $u = y$, $v = Ey$, and $D(y, Ey) = |A^T W A|$ be almost surely nonzero, then Ey has by the Implicit Function Theorem continuous partial derivatives of as many orders as are possessed by the function f_i in the neighborhood of (y, Ey), the unique point such that $f_i = 0$ for $i = 1, 2, \ldots, n$. From the definition of f_i, f_i has at least $p - 2$ continuous partial derivatives if p is an odd integer, and possesses all its continuous partial derivatives if p is even. These results lead to the following theorem.

Theorem 6. If the rank of the $m \times n$ matrix A is $n < m$, then almost surely for $1 < p < \infty$

   (i) $E \in C^\infty$ if p is an even integer;

   (ii) $E \in C^{p-2}$ if p is an odd integer;

   (iii) $E \in C^{[p]-2}$ if p is nonintegral where $[p]$ is the integral part of p, where C^k denotes the set of continuous functions with k continuous partial derivatives.

It is clear that if one replaces E with F and p with q that Theorem 6 remains true. By the definition of B and noting that A^g, a linear transformation, belongs to the class C^∞, it follows that B belongs to C^k where k is determined by p and q as required by Theorem 6. That is, if p and q are both even integers, then $k = \infty$; if p is an odd integer and q is an even integer, then $k = p$; if p and q are both odd integers then $k = \min[p, q]$. Conditions (i), (ii), and (iii) imply other similar values for k depending on the nature of p and q.

3. COMMON SOLUTIONS FOR n MATRIX EQUATIONS

Let n be a positive integer. Suppose A_i is a $p \times q$ matrix and B_i is a $p \times r$ matrix for $i = 1, 2, \ldots, n$.

Define $\quad C_1 = A_1,\, D_1 = B_1,\, E_1 = A_1^+ B_1,$ and $\quad F_1 = I - A_1^+ A_1$. Furthermore, define $C_k = A_k F_{k-1},\, D_k = B_k - A_k E_{k-1},\, E_k = E_{k-1} + F_{k-1} C_k^+ D_k,$ and $F_k = F_{k-1}(I - C_k^+ C_k)$ for $k = 2, 3, \ldots, n$.

Theorem 7. $A_i X = B_i$ for $i = 1, 2, \ldots, n$, has a common solution if and only if $C_i C_i^+ D_i = D_i$ for $i = 1, 2, \ldots, n$. The general common solution is $X = E_n + F_n Z$ where Z is arbitrary.

Proof. The proof is by mathematical induction. $A_1 X = B_1$ has a solution if and only if $A_1 A_1^+ B_1 = B_1$ or $C_1 C_1^+ D_1 = D_1$. The general solution is $X = A_1^+ B_1 + (I - A_1^+ A_1)Z = E_1 + F_1 Z$ where Z is arbitrary. Now assume $A_i X = B_i$ for $i = 1, 2, \ldots, k$ has a common solution if and only if $C_i C_i^+ D_i = D_i$ for $i = 1, 2, \ldots, k$. Also, assume the general common solution is $X = E_k + F_k Y$ where Y is arbitrary. $A_{k+1} X = B_{k+1}$ has a solution which is also a solution for $A_i X = B_i$ for $i = 1, 2, \ldots, k$ if and only if $B_{k+1} = A_{k+1}(E_k + F_k Y)$ or $A_{k+1} F_k Y = B_{k+1} - A_{k+1} E_k$ or $C_{k+1} Y = D_{k+1}$ has a solution for Y. But, $C_{k+1} Y = D_{k+1}$ has a solution if and only if $C_{k+1} C_{k+1}^+ D_{k+1} = D_{k+1}$. The general solution is $Y = C_{k+1}^+ D_{k+1} + (I - C_{k+1}^+ C_{k+1})Z$ where Z is arbitrary. Therefore $A_i X = B_i$ for $i = 1, 2, \ldots, k, k + 1$ has a common solution if and only if $C_i C_i^+ D_i = D_i$ for $i = 1, 2, \ldots, k, k + 1$. The general common solution is

$$X = E_k + F_k Y = E_k + F_k[C_{k+1}^+ D_{k+1} + (I - C_{k+1}^+ C_{k+1})Z]$$
$$= (E_k + F_k C_{k+1}^+ D_{k+1}) + [F_k(I - C_{k+1}^+ C_{k+1})]Z = E_{k+1} = F_{k+1} Z$$

where Z is arbitrary.

Corollary 7. Suppose A is a $p \times q$ matrix and B is a $p \times r$ matrix. For $i = 1, 2, \ldots, p$, define A_i to be the ith row of A, and B_i to be the ith row of B. $AX = B$ has a solution if and only if $C_i C_i^+ D_i = D_i$ for $i = 1, 2, \ldots, p$. The general solution is $X = E_p + F_p Z$ where Z is arbitrary.

Corollary 8. Suppose A_i is a $1 \times q$ vector for $i = 1, 2, \ldots, p$. If $A_1, A_2, \ldots, A_k$ are linearly independent, then $A_1, A_2, \ldots, A_k, A_{k+1}$ are linearly independent if and only if $C_{k+1} \neq 0$.

Corollary 9. Suppose A_i is a $1 \times q$ vector for $i = 1, 2, \ldots, p$, then $C_1, C_2, \ldots, C_p$ are orthogonal and span the space spanned by $A_1, A_2, \ldots, A_p$.

Corollary 10. Suppose A is a $p \times p$ matrix. For $i = 1, 2, \ldots, p$, define A_i to be the ith row of A, and B_i to be the ith $1 \times p$ unit vector. A is nonsingular if and only if $C_i \neq 0$ for $i = 1, 2, \ldots, p$. Furthermore, $A^{-1} = E_p$.

Corollary 11. Suppose A is a $p \times q$ matrix. For $i = 1, 2, \ldots, p$, define A_i to be the ith row of A, and B_i to be the scalar zero. X is a null vector of A if and only if there exists a $q \times 1$ vector Y such that $X = F_p Y$.

Corollary 12. Suppose P is a $p \times p$ matrix and λ is a complex number. Define $A = P - \lambda I$. For $i = 1, 2, \ldots, p$, define A_i to be the ith row of A, and B_i to be the scalar zero. λ is an eigenvalue of P if and only if $F_p \neq 0$. X is an eigenvector of P corresponding to the eigenvalue λ if and only if there exists a $p \times 1$ vector Y such that $X = F_p Y$.

Naturally, the treatment of Theorem 7 can be extended to include the common solutions for $XA_i = B_i$ for $i = 1, 2, \ldots, n$. Applications similar to those presented are obvious.

In all the corollaries the calculations of E_p and F_p involve matrix addition, matrix multiplication, and the determination of C_i^+ for $i = 1, 2, \ldots, p$. However, since the C_is are row matrices, the C_i^+s are easily obtained. If $C_i = 0$, then $C_i^+ = 0^T$ and $C_i C_i^+ = 0$. If $C_i \neq 0$, then $C_i^+ = \lambda^{-1} C_i^*$ and $C_i C_i^+ = 1$ where $\lambda = C_i C_i^*$. Furthermore, in some of the corollaries the D_is are scalars (sometimes zero).

EXERCISES

1. Let $Ax = b$ be an inconsistent system of $n + 1$ equations in n unknowns with $r(A) = n$. Let $C = I - AA^+$. Then show that the least squares residual d is given by $d = -Cb$. Also, show that

$$C = \frac{dd^*}{d^*d}.$$

2. Establish the properties listed in (7) through (12).
3. Prove Corollaries 3 through 6.
4. Prove Corollaries 7 through 12.
5. Determine whether it is possible to obtain a p-q generalized inverse for p and q strictly between 0 and 1.

6

APPLICATIONS

1. STATISTICAL APPLICATIONS

The most common application of generalized inverses of matrices is the solution of systems of equations. The general case has been discussed in Chapter 5. In this chapter selected application to specific problems is discussed. More applications can be found by consulting Appendix 2, the References, and recent literature.

1.1. Sequential Least Squares Parameter Estimation

In this section a sequential algorithm for least squares estimation of a parameter state vector is developed utilizing the properties of the pseudoinverse. This algorithm allows the estimation to begin after the first observation has been made, and requires no a priori knowledge of the initial state of the system. The problems of weighted least squares and deleting a bad observation are also considered. The problems associated with singular matrices encountered in iterative least squares procedures do not affect the algorithm.

The problem is to obtain a least squares solution for $\mathbf{x}$ in the equation $y = H\mathbf{x}$, where y is composed of t observation vectors, H is an $(n + k) \times p$ matrix, and $\mathbf{x}$ is a $p \times 1$ parameter state vector. The least squares solution of minimal Euclidean norm is given by

$$\hat{\mathbf{x}} = H^{+}y.$$

It will be assumed that n and k are multiples of r, the size of the observation vectors, that is, $n = n_1 r$, $k = n_2 r$.

A sequential method for computing the least squares estimate, after $n_1 + n_2$ observations have been made and without having to begin again from the beginning, is especially desirable in real time operation. This method allows one to move from the n_1th to the $(n_1 + n_2)$th least squares solution for the parameter state vector with a minimum of computations.

The problem of the matrix H^*H becoming singular does not affect this algorithm because the solution for x which has minimum Euclidean norm is chosen and the estimation procedure continues on. At the time the matrix H^*H becomes nonsingular this method gives the same solution as the conventional method. The method that will be outlined does not require the system to be fully determined before the estimation begins, nor does it require any a priori knowledge of the initial state of the system.

This section is divided into three major divisions. The first division exhibits the sequential algorithm. The second examines the problem of weighting. The third presents a method for deleting bad observations.

The Sequential Algorithm

Utilizing the properties of the pseudoinverse, an algorithm is developed that allows one to move sequentially from the n_1th to the $(n_1 + n_2)$th least squares solution for the parameter state vector with a minimum of operations.

Theorem 1. The pseudoinverse of any matrix

$$H = \begin{bmatrix} R \\ S \end{bmatrix}$$

can be written in the following form:

$$H^+ = (R^+ - TSR^+, T)$$

where $E = S(I - R^+R)$, $K = [I + (I - EE^+)SR^+R^{+*}S^*(I - EE^+)]^{-1}$, and $T = E^+ + (I - E^+S)R^+R^{+*}S^*K(I - EE^+)$.

Proof. This result is readily verified from Theorem 4 in Chapter 2.

Corollary 1. Let H_{n+k} be any $(n + k) \times p$ matrix partitioned as

$$H_{n+k} = \begin{bmatrix} H_n \\ H_k \end{bmatrix}$$

where H_n is $n \times p$ and H_k is $k \times p$, then

$$H_{n+k}^+ = (H_n^+ - T_k H_k H_n^+, T_k)$$

where

$$T_k = E_k^+ + (I - E_k^+ H_k)H_n^+ H_n^{+*}H_k^* K_k(I - E_k E_k^+),$$
$$K_k = [I + (I - E_k E_k^+)H_k H_n^+ H_n^{+*}H_k^*(I - E_k E_k^+)]^{-1},$$

and $E_k = H_k(I - H_n^+ H_n)$.

Proof. This result is immediate since H_{n+k} has the form specified in Theorem 1.

Let

$$y = \begin{pmatrix} y_1 \\ y_2 \\ \vdots \\ y_t \end{pmatrix}$$

where each y_i is an $r \times 1$ observation vector, then the least squares solution $\hat{x}_{n+k}$ of $y = Hx + e$ can be realized as a sequential process by writing H_{n+k}^+ and y in partitioned form as

$$\hat{x}_{n+k} = (H_n^+ - T_k H_k H_n^+, T_k)\begin{pmatrix} y_n \\ y_k \end{pmatrix}$$

where y_n consists of the first n_1 observation vectors and y_k consists of the last n_2 observation vectors. Multiplying gives

$$\hat{x}_{n+k} = H_n^+ y_n - T_k H_k H_n^+ y_n + T_k y_k,$$

and noting that $\hat{x}_n = H_n^+ y_n$ yields

$$\hat{x}_{n+k} = \hat{x}_n + T_k(y_k - H_k \hat{x}_n).$$

It should be noted that no a priori knowledge is necessary to begin the estimation procedure. One need only note that $\hat{x}_1 = H_1^+ y_1$ to start the procedure. In order to carry out this procedure it is only necessary to compute sequentially the two matrices $H_n^+ H_n$ and $H_n^+ H_n^{+*}$. Of these two matrices $H_n^+ H_n^{+*}$ can be shown to be the covariance matrix of $\hat{x}_n$.

To compute $H_n^+ H_n$ sequentially we have

$$H_{n+k}^+ H_{n+k} = (H_n^+ - T_k H_k H_n^+, T_k)\begin{pmatrix} H_n \\ H_k \end{pmatrix} = H_n^+ H_n + T_k H_k(I - H_n^+ H_n)$$

and to compute $H_n^+ H_n^{+*}$ sequentially we use the formula

$$H_{n+k}^+ H_{n+k}^{+*} = H_n^+ H_n^{+*} - H_n^+ H_k^{+*} H_k^* T_k^* - T_k H_k H_n^+ H_n^{+*}$$
$$+ T_k H_k H_n^+ H_n^{+*} H_k^* T_k^* + T_k T_k^*$$
$$= (I - T_k H_k)H_n^+ H_n^{+*}(I - T_k H_k)^* + T_k T_k^*.$$

Weighted Estimation

By weighted least squares we mean minimizing $(Hx - y)^* R^{-1}(Hx - y)$ where R is a hermitian definite matrix, hence there exists a matrix Q such that $Q^* Q = R^{-1}$. In order to do this it is only necessary to consider a matrix equation of the form $Qy = QHx + Qe$ instead of the equation $y = Hx + e$. This indicates that the least squares solution for x is given by

$$\hat{x} = (QH)^+ Qy.$$

It follows that the theoretical results are the same.

Deleting Observations

Suppose the tth observation has been made and it is then determined that the last n_2 observations are bad. It would be desirable to back up and delete these bad observations and then continue on in the estimation procedure without having to begin again from the beginning.

A method for deleting the last n_2 observations which are detected after the next estimate has been calculated will now be derived. This is done by partitioning the matrix H_{n+k} and the vector y as before so that $\hat{x}_{n+k}$ is given by

$$\hat{x}_{n+k} = \binom{H_n}{H_k}^+ \binom{y_n}{y_k}.$$

Partitioning H_{n+k}^+ as (F, W), then

$$\hat{x}_{n+k} = Fy_n + Wy_k. \tag{1}$$

The idea is to obtain $\hat{x}_n$ in terms of $\hat{x}_{n+k}$, H_k, y_k, and H_{n+k}^+ which are all available at the tth observation stage.

Working with Theorem 5 in Chapter 2, it is readily shown that

$$H_n^+ = (I - B_k B_k^+)[I + W(I - H_k W)^+ H_k]F \tag{2}$$

with $B_k^{+*} = W - W(I - H_k W)^+ (I - H_k W)$. Noting that $\hat{x}_n = H_n^+ y_n$ and using the expression for H_n^+ in (2) we have

$$\hat{x}_n = (I - B_k B_k^+)[I + W(I - H_k W)^+ H_k]Fy_n.$$

Replacing Fy_n by $\hat{x}_{n+k} - Wy_k$ we get

$$\hat{x}_n = (I - B_k B_k^+)[I + W(I - H_k W)^+ H_k](\hat{x}_{n+k} - Wy_k). \tag{3}$$

Since $(H_{n+k} H_{n+k})^+ H_{n+k}^* = H_{n+k}^+$ we find by partitioning H_{n+k} that

$$H_{n+k}^+ = C_{n+k}(H_n^*, H_k^*) \quad \text{with} \quad C_{n+k} = (H_{n+k}^* H_{n+k})^+,$$

which implies that $W = C_{n+k} H_k^*$. This substituted into (3) leads to

$$\hat{x}_n = (I - B_k B_k^+)[I + C_{n+k} H_k^*(I - H_k C_{n+k} H_k^*)^+ H_k](\hat{x}_{n+k} - C_{n+k} H_k^* y_k)$$

with $B_k = C_{n+k} H_k^* - C_{n+k} H_k^*(I - H_k C_{n+k} H_k^*)^+ (I - H_k C_{n+k} H_k^*).$

1.2. Distribution Theory

If x is a column vector of n random variables which have a joint n-dimensional Gaussian (or normal) distribution with mean vector m and covariance matrix V, we denote it as $x \sim N(m, V)$. In this, if $V = I$, then $y = \sum_{i=1}^{m} x_i^2 = x^T x$ has a known distribution called the noncentral chi-square, and this is written as $y \sim \chi^2(n, \lambda)$ where the so-called noncentrality parameter $\lambda = m^T m/2$. If $\lambda = 0$, the noncentral chi-square is the central chi-square.

Theorem 2. Let the $p \times 1$ random vector $x \sim N(0, V)$ where $r(V) = k \leq p$. A necessary and sufficient condition that a quadratic form $x^T A x$ has a chi-square distribution is that $V(AVA - A)V = 0$.

Proof. Let the $k \times 1$ random vector $y \sim N(0, I)$. Then x can be expressed as $x = By$ and $x^T A x$ can be expressed in terms of y as $y^T B^T A B y$. Since $y \sim N(0, I)$, a necessary and sufficient condition that $y^T B^T A B y$ has a chi-square distribution is that $B^T A B$ is idempotent, which is equivalent to $BB^T ABB^T ABB^T = BB^T ABB^T$, or $VAVAV = VAV$, which can be expressed as $V(AVA - A)V = 0$. The degrees of freedom of the chi-square is

$$\text{trace } B^T A B = \text{trace } ABB^T = \text{trace } AV.$$

Further results on the distribution of quadratic forms in singular normal variates can be found in Rayner and Livingstone.

We now proceed to establish formulas for the conditional means and covariances which are valid even when the joint distribution is singular.

Theorem 3. Let

$$\begin{bmatrix} x_1 \\ x_2 \end{bmatrix}$$

be a partitioned zero mean normal random vector with

$$S = \text{cov} \begin{bmatrix} x_1 \\ x_2 \end{bmatrix} = \begin{bmatrix} A & B \\ B^T & C \end{bmatrix},$$

$\text{cov}(x_1) = A$, and $\text{cov}(x_2) = C$, then the expected value of x_1, given $x_2 = b$, and the covariance matrix of x_1, given $x_2 = b$, are given by $E(x_1 \mid x_2 = b) = BC^+ b$ and $\text{cov}(x_1 \mid x_2 = b) = A - BC^+ B^T$.

Proof. We will derive the formulas for the conditional mean and covariance of x_1, given $x_2 = b$, by representing x_1 in such a way that it is obvious what conditioning on x_2 means. We need only the rule for computing covariances under a linear transformation, that is, if y has covariance matrix S, then My has covariance matrix MSM^T. Let $y = x_1 - BC^+ x_2$. Then the elements of the random vector y have zero means, and the covariance matrix of the composite vector

$$\begin{bmatrix} y \\ \mathbf{x}_2 \end{bmatrix} = \begin{bmatrix} I & -BC^+ \\ 0 & I \end{bmatrix}\begin{bmatrix} \mathbf{x}_1 \\ \mathbf{x}_2 \end{bmatrix}$$

is

$$\begin{bmatrix} I & -BC^+ \\ 0 & I \end{bmatrix}\begin{bmatrix} A & B \\ B^T & C \end{bmatrix}\begin{bmatrix} I & 0 \\ -C^+B^T & I \end{bmatrix} = \begin{bmatrix} A - BC^+B^T & B - BC^+C \\ B^T - CC^+B^T & C \end{bmatrix}.$$

To establish that the off-diagonal blocks are 0, the general covariance matrix V is positive semidefinite. Hence there exists a matrix P such that $V = P^T P$. Partitioning, we get

$$V = \begin{bmatrix} P_1^{\ T} \\ P_2^{\ T} \end{bmatrix}(P_1, P_2) = \begin{bmatrix} P_1^{\ T}P_1 & P_1^{\ T}P_2 \\ P_2^{\ T}P_1 & P_2^{\ T}P_2 \end{bmatrix},$$

and the column space of $P_2^{\ T}P_1$ lies in the column space of $P_2^{\ T}$ which is the same as the column space of $P_2^{\ T}P_2$. Hence, without loss of generality, we find that the cloumns of B^T lie in the column space of C. But, in that case $CC^+B^T = B^T$ since CC^+ is the projector of the column space of C. Hence $B^T - CC^+B^T = 0$ implies that $B - BC^+C = 0$. Thus the covariance matrix becomes

$$\text{cov}\begin{bmatrix} y \\ \mathbf{x}_2 \end{bmatrix} = \begin{bmatrix} A - BC^+B^T & 0 \\ 0 & C \end{bmatrix}.$$

Hence the covariance matrix of y is $A - BC^+B^T$, and y is independent of $\mathbf{x}_2$. Because of this independence, it follows immediately that the conditional distribution of $\mathbf{x}_1 = y + BC^+\mathbf{x}_2$, given $\mathbf{x}_2 = b$, is normal with mean BC^+b and covariance that of y.

It should be noted that these formulas for conditional mean and covariance apply not only for the normal, but for any joint distribution for which zero correlation implies statistical independence.

1.3. The Independence of Quadratic Forms in Normal Variates

F. A. Graybill [1] has shown that if $y \sim N(\mu, I)$, a necessary and sufficient condition for the independence of $y^T Ay$ and $y^T By$, where A and B are symmetric and positive semidefinite, is $AB = 0$.

The purpose of this sectionis to extend these results to more general situations and to establish sufficient conditions for independence in the general case.

We begin by stating some lemmas which will be useful at a later point.

Lemma 1. Let

$$A = \begin{pmatrix} A_{11} & A_{12} \\ A_{21} & A_{22} \end{pmatrix}$$

be an $n \times n$ symmetric positive semidefinite matrix partitioned so that A_{11} is $m \times m$. The row null space of A_{11} is a subspace of the row null space of A_{12}, and the row null space of A_{22} is a subspace of the row null space of A_{21}. Similar relationships hold between the column null spaces of A_{11} and A_{21} and between the column null spaces of A_{22} and A_{12}.

Proof. We shall prove only that the row null space of A_{11} is a subspace of the row null space of A_{12}. The other relationships follow using similar arguments.

Since A is positive semidefinite, there exists an $n \times n$ matrix T such that $A = T^T T$. Let

$$T = \begin{pmatrix} T_{11} & T_{12} \\ T_{21} & T_{22} \end{pmatrix}$$

be partitioned in the same way as A. Suppose X_1 is an $m \times 1$ vector such that $X_1 A_{11} = 0$. It is clear that the $n \times 1$ vector

$$X = \begin{pmatrix} X_1 \\ 0 \end{pmatrix}$$

satisfies $X^T A X = 0$ so that $X^T T^T = 0$. Thus $X_1{}^T T_{11}{}^T = 0$ and $X_1{}^T T_{21}{}^T = 0$. Now, since $A_{12} = T_{11}{}^T T_{12} + T_{21}{}^T T_{22}$, we have $X_1{}^T A_{12} = 0$.

Lemma 2. Let

$$A = \begin{pmatrix} A_{11} & A_{12} \\ A_{21} & A_{22} \end{pmatrix}$$

be an $n \times n$ symmetric positive semidefinite matrix partitioned so that A_{11} is $m \times m$. Then $A_{11} A_{11}^+ A_{12} = A_{12}$.

Proof. Consider the matrix $I - A_{11} A_{11}^+$. Since $(I - A_{11} A_{11}^+) A_{11} = 0$, the rows of $I - A_{11} A_{11}^+$ are in the row null space of A_{11}. Thus, by Lemma 1, $(I - A_{11} A_{11}^+) A_{12} = 0$ so that $A_{12} = A_{11} A_{11}^+ A_{12}$.

Theorem 4. Let

$$y \sim N\left[\begin{pmatrix} \mu_1 \\ \mu_2 \end{pmatrix}, \begin{pmatrix} I_r & 0 \\ 0 & 0 \end{pmatrix} \right]$$

where μ_1 is $r \times 1$. Let

$$A = \begin{pmatrix} A_{11} & A_{12} \\ A_{21} & A_{22} \end{pmatrix} \quad \text{and} \quad B = \begin{pmatrix} B_{11} & B_{12} \\ B_{21} & B_{22} \end{pmatrix},$$

where A_{11} and B_{11} are $r \times r$, be symmetric, and let A be positive semidefinite. A necessary and sufficient condition for the independence of $y^T A y$ and $y^T B y$ is $A_{11} B_{11} = 0$ and

$$A \begin{pmatrix} I_r & 0 \\ 0 & 0 \end{pmatrix} B \begin{pmatrix} \mu_1 \\ \mu_2 \end{pmatrix} = 0.$$

Proof of Sufficiency. First we observe that if y^T is partitioned as (y_1^T, y_2^T), where y_1 is $r \times 1$, $y_2 = \mu_2$ with probability 1 since $\text{var}(y_2) = 0$. Then $y^T A y$ and $y^T B y$ can be written as

$$y^T A y = y_1^T A_{11} y_1 + 2 y_1^T A_{12} \mu_2 + \mu_2^T A_{22} \mu_2, \tag{4}$$

$$y^T B y = y_1^T B_{11} y_1 + 2 y_1^T B_{12} \mu_2 + \mu_2^T B_{22} \mu_2. \tag{5}$$

Since $A_{11} B_{11} = 0$ there exists an orthogonal matrix P_{11} such that $P_{11}^T A_{11} P_{11} = D_1$ and $P_{11}^T B_{11} P_{11} = D_2$, where D_1 and D_2 are diagonal and D_1 is positive semidefinite. We also observe that $D_1 D_2 = 0$.

Letting

$$P = \begin{pmatrix} P_{11} & 0 \\ 0 & I \end{pmatrix}$$

and $z = P^T y$ we have

$$z \sim N\left[P^T \mu, \begin{pmatrix} I_r & 0 \\ 0 & 0 \end{pmatrix} \right],$$

and (4) and (5) become

$$y^T A y = z_1^T D_1 z_1 + 2 z_1^T P_{11} A_{12} \mu_2 + \mu_2^T A_{22} \mu_2, \tag{6}$$

$$y^T B y = z_1^T D_2 z_1 + 2 z_1^T P_{11} B_{12} \mu_2 + \mu_2^T B_{22} \mu_2. \tag{7}$$

It is clear that z_i appears in $z_1^T D_1 z_1$ iff $E^T(i) D_1 \neq 0$ where $E(i)$ is the $r \times 1$ vector with 1 in the ith row and zeros elsewhere. Also, by Lemma 1, $E^T(i) D_1 = 0$ implies that $E^T(i) P_{11}^T A_{12} = 0$. Thus z_i appears in (6) iff $E^T(i) D_1 \neq 0$.

Suppose $E^T(i) D_1 \neq 0$. Then, since $D_1 D_2 = 0$, $E^T(i) D_2 = 0$. Also, since

$$A \begin{pmatrix} I_r & 0 \\ 0 & 0 \end{pmatrix} B \begin{pmatrix} 0 \\ \mu_2 \end{pmatrix} = 0,$$

we have $A_{11} B_{12} \mu_2 = 0$ so that $D_1 P_{11}^T B_{12} \mu_2 = 0$ and therefore $E(i) P_{11}^T B_{12} \mu_2 = 0$. Thus, if z_i appears in (6), z_i does not appear in (7).

Since $z_1 \sim N(P_{11}\mu_1, I_r)$, the z_is are jointly independent, so that $y^T A y$ and $y^T B y$ can be written as functions of disjoint subsets of a set of independent random variables. Thus $y^T A y$ and $y^T B y$ are independent.

Proof of Necessity. Again we write $y^T A y$ and $y^T B y$ as in (4) and (5). Since A_{11} is positive semidefinite, there exists an orthogonal P_{11} such that $P_{11}^T A_{11} P_{11} = D$ where D is diagonal and positive semidefinite. Let

$$P = \begin{pmatrix} P_{11} & 0 \\ 0 & I \end{pmatrix}$$

and $z = P^T y$. Then (5) becomes

$$y^T B y = z_1^T P_{11}^T B_{11} P_{11} z_1 + 2 z_1^T P_{11T} B_{12} \mu_2 + \mu_2^T B_{22} \mu_2 \tag{8}$$

and applying Lemma 2, (4) becomes

$$
\begin{aligned}
y^T A y &= z_1{}^T D z_1 + 2z_1{}^T P_{11}{}^T A_{12} \mu_2 + \mu_2{}^T A_{22} \mu_2 \\
&= (z_1 + D^+ P_{11}{}^T A_{12} \mu_2)^T D(z_1 + D^+ P_{11}{}^T A_{12} \mu_2) \\
&\quad + \mu_2{}^T (A_{22} - A_{21} A_{11}^+ A_{12}) \mu_2 .
\end{aligned} \tag{9}
$$

To simplify notation we let $w = z_1 + D^+ P_{11}{}^T A_{12} \mu_2$,

$$
\alpha = \mu_2{}^T (A_{22} - A_{21} A_{11}^+ A_{12}) \mu_2, \lambda = D^+ P_{11}{}^T A_{12} \mu_2,
$$

and

$$
(w - \lambda)^T P_{11}{}^T B_{11} P_{11}(w - \lambda) + 2(w - \lambda)^T P_{11}{}^T B_{12} \mu_2 = Q_B(w).
$$

Thus

$$
\begin{pmatrix} A_{11} & A_{12} \\ A_{21} & A_{22} \end{pmatrix} \begin{pmatrix} I_r & 0 \\ 0 & 0 \end{pmatrix} \begin{pmatrix} B_{11} & B_{12} \\ B_{21} & B_{22} \end{pmatrix} \begin{pmatrix} \mu_1 \\ \mu_2 \end{pmatrix} = 0.
$$

Theorem 4 is easily extended to the case of a general covariance matrix by the following.

Theorem 5. Let $y \sim N(\mu, V)$ where rank $(V) = r$. Let A and B be symmetric, and let A be positive semidefinite. A necessary and sufficient condition for the independence of $y^T A y$ and $y^T B y$ is $AVBV = 0$ and $AVB\mu = 0$.

Proof. Since V is symmetric positive semidefinite of rank r, there exists nonsingular Q such that

$$
Q^T V Q = \begin{pmatrix} I_r & 0 \\ 0 & 0 \end{pmatrix}
$$

Let $z = Q^T y$. Then

$$
z \sim N\left[Q^T \mu, \begin{pmatrix} I_r & 0 \\ 0 & 0 \end{pmatrix} \right]
$$

and

$$
\begin{aligned}
y^T A y &= z^T Q^{-1} A (Q^T)^{-1} z, \\
y^T B y &= z^T Q^{-1} B (Q^T)^{-1} z.
\end{aligned}
$$

Applying Theorem 4, $y^T A y$ and $y^T B y$ are independent iff

$$
Q^{-1} A (Q^T)^{-1} \begin{pmatrix} I_r & 0 \\ 0 & 0 \end{pmatrix} Q^{-1} B (Q^T)^{-1} \begin{pmatrix} I_r & 0 \\ 0 & 0 \end{pmatrix} = 0 \tag{10}
$$

and

$$
Q^{-1} A (Q^T)^{-1} \begin{pmatrix} I_r & 0 \\ 0 & 0 \end{pmatrix} Q^{-1} B (Q^T)^{-1} Q^T \mu = 0. \tag{11}
$$

But (10) is true iff $AVBV = 0$, and (11) is true iff $AVB\mu = 0$.

Corollary 2. If $y^T A y$ and $y^T B y$ are independent, where $y \sim N(\mu, V)$, A and B are symmetric, and A is positive semidefinite, then $y^T A y$ and $y^T B y$ can be written by a linear transformation of y as functions of disjoint subsets of a set of independent random variables.

Corollary 3. If $y \sim N(\mu, V)$ and A and B are symmetric and positive semidefinite, then $y^T A y$ and $y^T B y$ are independent iff $A V B = 0$.

Proof. If B is positive semidefinite, $A V B V = 0$ iff $A V B = 0$.

Corollary 4. If $y \sim N(\mu, V)$, V is nonsingular, A and B are symmetric, and A is positive semidefinite, then $y^T A y$ and $y^T B y$ are independent iff $A V B = 0$.

Also of interest in many instances is the independence of certain quadratic and linear forms. The following theorem establishes conditions under which this type of independence can occur.

Theorem 6. Let $y \sim N(\mu, V)$ where rank $(V) = r$ and let B be a symmetric matrix. Then Ay and $y^T B y$ are independent iff $A V B V = 0$ and $A V B \mu = 0$.

Proof. First suppose that Ay and $y^T B y$ are independent. Then it is clear that $y^T A^T A y$ and $y^T B y$ are independent. Applying Theorem 5 we have $A^T A V B V = 0$ and $A^T A V B \mu = 0$. Thus $A V B V = 0$ and $A V B \mu = 0$.

Now suppose that $A V B V = 0$ and $A V B \mu = 0$. Then $A^T A V B V = 0$ and $A^T A V B \mu = 0$ so that by Theorem 5, $y^T A^T A y$ and $y^T B y$ are independent. By Corollary 2, there exists a linear transformation $y = Qz$ such that Q is nonsingular,

$$z \sim N\left[v, \begin{pmatrix} I_r & 0 \\ 0 & 0 \end{pmatrix} \right]$$

and $z^T Q^T A^T A Q z$ and $Z^T Q^T B Q z$ are functions of disjoint subsets of a set of independent random variables. But $A Q z$ is a function of the same variables as $(A Q z)^T (A Q z)$. Thus $A Q z$ and $z^T Q^T B Q z$ are functions of disjoint subsets of a set of independent random variables so that Ay and $y^T B y$ are independent.

Corollary 5. If $y \sim N(\mu, V)$ and B is symmetric positive semidefinite, then Ay and $y^T B y$ are independent iff $A V B = 0$.

Proof. If B is positive semidefinite, $A V B V = 0$ iff $A V B = 0$.

Corollary 6. If $y \sim N(\mu, V)$, V is nonsingular, and B is symmetric, then Ay and $y^T B y$ are independent iff $A V B = 0$.

Theorem 7. Let

$$y \sim N\left[\begin{pmatrix} \mu_1 \\ \mu_2 \end{pmatrix}, \begin{pmatrix} I_r & 0 \\ 0 & 0 \end{pmatrix} \right]$$

where μ_1 is $r \times 1$. Let

$$A = \begin{pmatrix} A_{11} & A_{12} \\ A_{21} & A_{22} \end{pmatrix} \quad \text{and} \quad B = \begin{pmatrix} B_{11} & B_{12} \\ B_{21} & B_{22} \end{pmatrix},$$

where A_{11} and B_{11} are $r \times r$, be symmetric. A sufficient condition for the independence of $y^T A y$ and $y^T B y$ is $A_{11} B_{11} = 0$, $A_{11} B_{12} \mu_2 = 0$, $\mu_2^T A_{21} B_{11} = 0$, and $\mu_2^T A_{21} B_{12} \mu_2 = 0$.

Proof. Partitioning y and selecting an orthogonal matrix P as in the sufficiency proof of Theorem 4, we again write

$$y^T A y = z_1^T D_1 z_1 + 2 z_1^T P_{11}^T A_{12} \mu_2 + \mu_2^T A_{22} \mu_2, \tag{10}$$

$$y^T B y = z_1^T D_2 z_1 + 2 z_1^T P_{11}^T B_{12} \mu_2 + \mu_2^T B_{22} \mu_2 \tag{13}$$

where $z_1 \sim N(P_{11}^T \mu_1, I)$, D_1 and D_2 are diagonal, and $D_1 D_2 = 0$.

Suppose z_i appears in $z_1^T D_1 z_1$. Then, since $D_1 D_2 = 0$, z_i does not appear in $z_1^T D_2 z_1$. Also, since $A_{11} B_{12} \mu_2 = 0$, $D_1 P_{11}^T B_{12} \mu_2 = 0$ so that the ith row of $P_{11}^T B_{12} \mu_2$ is zero. Thus z_i does not appear in $z_1^T P_{11}^T B_{12} \mu_2$.

Now suppose z_i appears in $z_1^T P_{11}^T A_{12} \mu_2$. Since $B_{11} A_{12} \mu_2 = 0$, $D_2 P_{11}^T A_{12} \mu_2 = 0$ so that the ith row of D_2 is zero. Thus z_i does not appear in $z_1^T D_2 \mu_1$. Also, since $\mu_2^T A_{21} P_{11} P_{11}^T B_{12} \mu_2 = \mu_2^T A_{21} B_{12} = 0$, the ith row of $P_{11}^T B_{12} \mu_2$ is zero. Therefore z_i does not appear in $z_1^T P_{11}^T B_{12} \mu_2$.

We have shown that if z_i appears in $z_1^T D_1 z_1 + 2 z_1^T P_{11}^T A_{12} \mu_2 + \mu_2^T A_{22} \mu_2$, z_i does not appear in $z_1^T D_2 z_1 + 2 z_1^T P_{11}^T B_{12} \mu_2 + \mu_2^T B_{22} \mu_2$. Thus $y^T A y$ and $y^T B y$ can be expressed as functions of disjoint subsets of a set of independent random variables and are therefore independent.

It will be useful in Theorem 8 to note that $A_{11} B_{11} = 0$, $A_{11} B_{12} \mu_2 = 0$, $\mu_2^T A_{21} B_{11} = 0$, and $\mu_2^T A_{21} B_{12} \mu_2 = 0$ are equivalent to

$$\begin{pmatrix} I_r & 0 \\ 0 & 0 \end{pmatrix} A \begin{pmatrix} I_r & 0 \\ 0 & 0 \end{pmatrix} B \begin{pmatrix} I_r & 0 \\ 0 & 0 \end{pmatrix} = 0, \quad \begin{pmatrix} I_r & 0 \\ 0 & 0 \end{pmatrix} A \begin{pmatrix} I_r & 0 \\ 0 & 0 \end{pmatrix} B \mu = 0,$$

$$\mu^T A \begin{pmatrix} I_r & 0 \\ 0 & 0 \end{pmatrix} B \begin{pmatrix} I_r & 0 \\ 0 & 0 \end{pmatrix} = 0, \quad \text{and} \quad \mu^T A \begin{pmatrix} I_r & 0 \\ 0 & 0 \end{pmatrix} B \mu = 0.$$

Theorem 8. Let $y \sim N(\mu, V)$ and let A and B be symmetric. A sufficient condition for the independence of $y^T A y$ and $y^T B y$ is $VAVBV = 0$, $VAVB\mu = 0$, $\mu^T AVBV = 0$, and $\mu^T AVB\mu = 0$.

Proof. Since V is symmetric and positive semidefinite, there exists a nonsingular matrix Q such that

$$Q^T V Q = \begin{pmatrix} I_r & 0 \\ 0 & 0 \end{pmatrix}$$

where r is the rank of V. Let $z = Q^T y$, then

$$z \sim N\left(Q^T \mu, \begin{pmatrix} I_r & 0 \\ 0 & 0 \end{pmatrix}\right),$$

$y^T A y = z^T Q^{-1} A (Q^T)^{-1} z$, and $y^T B y = z^T Q^{-1} B (Q^T)^{-1} z$.

Applying the results of Theorem 7 we see that $y^T A y$ and $y^T B y$ are independent if $VAVBV = 0$, $VABV\mu = 0$, $\mu^T AVBV = 0$, and $\mu^T AVB\mu = 0$.

Theorem 9. Let $y \sim N(\mu, V)$ and let A and B be symmetric. A sufficient condition for the independence of $y^T A y + 2y^T \alpha + a$ and $y^T B y + 2y^T \beta + b$ is $VAVBV = 0$, $VAVB\beta = 0$, $\alpha AVBV = 0$, and $\alpha AVB\beta = 0$.

Proof. Choose matrices A_{12} and B_{12}, of the same dimension, and a vector μ_2 such that $A_{12}\mu_2 = \alpha$ and $B_{12}\mu_2 = \beta$. Let

$$z = \begin{pmatrix} y \\ \mu_2 \end{pmatrix}, \quad \hat{A} = \begin{pmatrix} A & A_{12} \\ A_{12}{}^T & 0 \end{pmatrix}, \quad \text{and} \quad \hat{B} = \begin{pmatrix} B & B_{12} \\ B_{21}{}^T & 0 \end{pmatrix}.$$

Then

$$z \sim N\left[\begin{pmatrix} \mu \\ \mu_2 \end{pmatrix}, \begin{pmatrix} V & 0 \\ 0 & 0 \end{pmatrix}\right],$$

$z^T \hat{A} z = y^T A y + 2y^T \alpha$, and $z^T \hat{B} z = y^T B y + 2y^T \beta$. Now by Theorem 8, $z^T \hat{A} z$ and $z^T \hat{B} z$ are independent if

$$\begin{pmatrix} V & 0 \\ 0 & 0 \end{pmatrix} \hat{A} \begin{pmatrix} V & 0 \\ 0 & 0 \end{pmatrix} \hat{B} \begin{pmatrix} V & 0 \\ 0 & 0 \end{pmatrix} = 0, \quad \begin{pmatrix} V & 0 \\ 0 & 0 \end{pmatrix} \hat{A} \begin{pmatrix} V & 0 \\ 0 & 0 \end{pmatrix} \hat{B} \begin{pmatrix} \mu \\ \mu_2 \end{pmatrix} = 0,$$

$$\begin{pmatrix} \mu \\ \mu_2 \end{pmatrix}^T \hat{A} \begin{pmatrix} V & 0 \\ 0 & 0 \end{pmatrix} \hat{B} \begin{pmatrix} V & 0 \\ 0 & 0 \end{pmatrix} = 0, \quad \text{and} \quad \begin{pmatrix} \mu \\ \mu_2 \end{pmatrix}^T \hat{A} \begin{pmatrix} V & 0 \\ 0 & 0 \end{pmatrix} \hat{B} \begin{pmatrix} \mu \\ \mu_2 \end{pmatrix} = 0.$$

These conditions are equivalent to

$$VAVBV = 0, \quad VAVB\beta = 0, \quad \alpha AVBV = 0, \quad \text{and} \quad \alpha AVB\beta = 0.$$

Note that $y^T A y$, where $y \sim N(\mu, V)$, can be expressed as

$$y^T A y = (y - \mu)^T A (y - \mu) + 2(y - \mu)^T A\mu - \mu^T A\mu$$

so that one need only consider quadratic expressions where the variable, for example, **x**, has 0 mean. For the independence conditions on the matrices in that form, one should consult the paper by Good along with the corrections presented in the paper by Shanbhag.

1.4. The Fixed Point Probability Vector of Regular or Ergodic Transition Matrices

In what follows, α, β, and η will denote $1 \times p$ vectors, P will denote either a regular or an ergodic $p \times p$ transition matrix, and 1^T will denote the $1 \times p$ row vector with every component unity.

It is convenient to include two lemmas, the first of which is a well-known result in the theory of finite Markov chains.

Lemma 3. If P is a regular or an ergodic transition matrix, then there exists a unique $1 \times p$ probability vector α such that $\alpha P = \alpha$. If P is regular

$$P^l \to \xi\alpha \quad \text{as} \quad l \to \infty.$$

If P is ergodic, then for $0 < \delta < 1$

$$\sum_{i=0}^{l} \binom{l}{i} \delta^{l-i}(1-\delta)^i P^i \to \xi\alpha \quad \text{as} \quad l \to \infty.$$

Lemma 4. If P is a regular or an ergodic transition matrix and α is the unique probability vector such that $\alpha P = \alpha$, and if $\eta P = \eta$ for some $1 \times p$ vector η, then $\eta = k\alpha$ for some scalar k.

Proof. From Lemma 3 it is easy to see, in the regular case, that $\eta = \eta\xi\alpha = k\alpha$ since $\eta\xi$ is a scalar. The ergodic case follows from the fact that if $0 < \delta < 1$ then $[\delta I + (1 - \delta)P]$ is regular.

We will now use these lemmas to prove the following.

Theorem 10. If P is a regular or an ergodic transition matrix and

$$\beta = \frac{1^T[I - (P - I)(P - I)^+]}{1^T[I - (P - I)(P - I)^+]1},$$

then β is the unique probability vector such that $\beta P = \beta$ (i.e., $\beta = \alpha$).

Proof. We will first show that the scalar $d = 1^T[I - (P - I)(P - I)^+]1$, in our definition of β, is nonzero. To do this note that $(P - I)(P - I)^+$ is idempotent and symmetric so that $d = 1^T[I - (P - I)(P - I)^+]1 = \{[I - (P - I)(P - I)^+]1\}^T\{[I - (P - I)(P - I)^+]1\}$. Hence, if $d = 0$, then $[I - (P - I)(P - I)^+]1 = 0$. However, this would imply (since, for the α in Lemma 3, $\alpha P = \alpha$ and $\alpha 1 = 1$) that $1 = \alpha 1 = \alpha[(P - I)(P - I)^+]1 = 0$, a contradiction.

It follows from Lemma 4 and the equations $\beta P = \beta$ and $\beta 1 = 1$ that $\beta = \alpha$.

The given form of the fixed point probability vector does not require the calculation of powers of P nor is it dependent upon a limiting process. Moreover, the form is independent of any distinction between the regular and ergodic case. It is only necessary to calculate $(P - I)^+$. In fact, the proof of the theorem only requires that $(P - I)^+$ be chosen in such a manner that $(P - I)(P - I)^+$ be symmetric and $(P - I)(P - I)^+(P - I) = (P - I)$. Hence, in the theorem, we may replace $(P - I)^+$ by any solution of only two of the equations defining the pseudoinverse, namely, any solution X such that $(P - I)X(P - I) = (P - I)$ and $[(P - I)X]^T = (P - I)X$. It might also be noted that the scalar $1^T[I - (P - I)(P - I)^+]1$ is the sum of the elements of the matrix $[I - (P - I)(P - I)^+]$, and the components of the vector $1^T[I - (P - I)(P - I)^+]$ are simply the column sums of the matrix $[I - (P - I)(P - I)^+]$.

1.5. Stochastic Matrices

In this section we give an application of a spectral inverse to stochastic matrices where the spectral property of the generalized inverse plays a very important role.

Let A be a stochastic matrix, that is, $A \geq 0$ and $A1 = 1$. A matrix A is said to be reducible if and only if there is a permutation matrix P such that

$$PAP^* = \begin{bmatrix} B & 0 \\ C & D \end{bmatrix}$$

where B and D are square matrices. Otherwise, the matrix A is called irreducible. For any reducible matrix there is a permutation matrix P such that

$$PAP^* = \begin{bmatrix} A_1 & 0 & \cdots & 0 & 0 & \cdots & 0 \\ 0 & A_2 & & 0 & 0 & \cdots & 0 \\ & & \cdots\cdots\cdots\cdots\cdots\cdots\cdots\cdots\cdots & & & & \\ 0 & 0 & & A_k & 0 & \cdots & 0 \\ A_{k+1,1} & A_{k+1,2} & & A_{k+1,k} & A_{k+1} & \cdots & 0 \\ & & \cdots\cdots\cdots\cdots\cdots\cdots\cdots\cdots\cdots & & & & \\ A_{n,1} & A_{n,2} & \cdots & A_{n,k} & A_{n,k+1} & \cdots & A_n \end{bmatrix} \qquad (14)$$

where the A_i, $i = 1, 2, \ldots, n$ are irreducible. We say that A is completely reducible if and only if there is a permutation matrix P such that $PAP^* = \text{diag}(A_1, A_2, \ldots, A_n)$.

Theorem 11. Let A be a stochastic matrix. Necessary and sufficient conditions for a spectral inverse A^s to be stochastic are that A be either completely reducible or irreducible and every nonzero eigenvalue of A lie on the unit circle.

Proof. We consider the necessity of the conditions. It is known that the eigenvalues of a stochastic matrix lie in the closed unit disc. Consequently, it follows from the definition of A^s that if A^s is stochastic, then all nonzero eigenvalues of A (and A^s) must lie on the unit circle in the complex plane.

Let A be reducible. Then there exists a permutation matrix P such that $\hat{A} = PAP^*$ where $\hat{A}$ has the form (14). Since P is a permutation matrix, $PP^* = I$. Thus $\hat{A}$ and A are similar, hence they have the same eigenvalues. Because of the triangular form of $\hat{A}$, the eigenvalues of A are precisely those of all of the A_i, $i = 1, 2, \ldots, n$. Suppose that there is an i greater than k such that not all of the $A_{i,1}, A_{i,2}, \ldots, A_{i,i-1}$ are zero. But in this case, the spectral radius of A_i is less than the spectral radius of A (Gantmacher [1], p. 92). Thus the eigenvalues of A_i are in modulus less than 1. This is a contradiction. Hence A is completely reducible. Thus the proof of the necessity is concluded.

The following lemma will be needed in the proof of the sufficiency.

Lemma 5. If A is stochastic, irreducible, and has all of its nonzero eigenvalues on the unit circle, then 0 is a simple root of the minimum equation of A.

Proof. It follows from a well-known result owing to Frobenius (Gantmacher [1], p. 65], that the minimum equation for A is of the form $x^p(x^h - 1) = 0$. Factoring the polynomial in this equation,

$$x^p(x^h - 1) = (x - 1)(x^{p+h-1} + x^{p+h-2} + \cdots + x^p). \tag{15}$$

Thus

$$(A - I)(A^{p+h-1} + A^{p+h-2} + \cdots + A^p) = 0. \tag{16}$$

Now a Jordan form for A is

$$C = \begin{bmatrix} 1 & 0 & 0 & \cdots & 0 & 0 \\ 0 & \omega_1 & 0 & \cdots & 0 & 0 \\ 0 & 0 & \omega_2 & \cdots & 0 & 0 \\ \cdots & \cdots & \cdots & \cdots & \cdots & \cdots \\ 0 & 0 & 0 & \cdots & \omega_{h-1} & 0 \\ 0 & 0 & 0 & \cdots & 0 & N \end{bmatrix}$$

where $\omega_i, i = 1, 2, \ldots, h - 1$, are the hth roots of unity different from 1, and N is a square matrix whose elements are all zero except for the diagonal above the principal diagonal. Then

$$A^{p+h-1} + A^{p+h-2} + \cdots + A^p = P(C^{p+h-1} + \cdots + C^p)P^{-1}$$
$$= P[\text{diag}(h, 0, 0, \ldots 0)]P^{-1}. \tag{17}$$

Thus the ith row of the sum of the matrices on the left in (17) is $hp_{i1}P_1^{-1}$ where P_1^{-1} is the first row of P^{-1}. But each matrix in the sum on the left in (17) is a stochastic matrix. Hence the sum of the elements in a given row of the sum is h. Now the first column of P is an eigenvector of A corresponding to 1. Hence we may take $p_{i1} = 1$, for $i = 1, 2, \ldots, n$. Thus the sum of the elements of the first row of P^{-1} is 1. Now if $p > 1$, consider the sum

$$A^{p+h-2} + A^{p+h-3} + \cdots + A^{p-1} = PDP^{-1} \tag{18}$$

where $d_{11} = h, d_{h+1,h+p} = 1$, and all other elements of D are zero. Then the ith row of PDP^{-1} is $hp_{i1}P_1^{-1} + P_{i,h+1}P_{h+p}^{-1}$. The summands on the left in (18) are each stochastic. Thus the sum of each row of PDP^{-1} must be h.

But $p_{i, h+1}$ is not zero for at least one value of i between 1 and n, and the row P_{h+p}^{-1} is not identically zero. Thus we have contradicted the fact that the sum of the elements of P_1^{-1} is 1. Hence $p \le 1$.

Returning to the proof of the sufficiency portion of Theorem 11, suppose that A is completely reducible and all of the nonzero eigenvalues of A lie on the unit circle. Then there is a permutation matrix P such that $PAP^* = \text{diag}(A_1, A_2, \ldots, A_g)$. Since each of the A_i is irreducible and the nonzero eigenvalues of each of the A_i lie on the unit circle, the nonzero eigenvalues of A_i are precisely the h_ith roots of unity. Thus it follows that A satisfies $A^P(A^k - I)$ where $k = \text{l.c.m.}\ (h_1, h_2, \ldots, h_g)$. From Lemma 5, either $p = 0$ or $p = 1$. If $p = 0$, then A is nonsingular and $A^s = A^{-1} = A^{k-1}$. If $p = 1$, then A is singular and $A^s = A^{k-1}$. In either case, A^s is stochastic.

2. OTHER APPLICATIONS

In this section selected applications are given that do not require any knowledge of statistics.

2.1. The Algebraic Eigenvalue Problem

We are concerned here with the problem of finding the scalars λ and the vectors x which satisfy the equation

$$Ax = \lambda Bx. \tag{19}$$

If B is nonsingular, (19) is reduced to the ordinary algebraic eigenproblem

$$Cx = \lambda x \tag{20}$$

by taking $C = B^{-1}A$.

If B is singular, let $B = P^{-1}JP$ where J is the Jordan canonical form of B. The matrix $X = P^{-1}J^+P$ is readily shown to be a spectral inverse of B. See problems 9 and 10 at the end of Chapter 3.

Lemma 6. If x is a solution of equation (19) corresponding to λ, then $y = Px$ where $B = P^{-1}JP$, with λ satisfying

$$PAP^{-1}y = \lambda PBP^{-1}y \tag{21}$$

and conversely, if λ and y represent a solution of (21), then λ and $x = P^{-1}y$ satisfy (19).

Proof. Premultiplying (19) by P and inserting $P^{-1}P$ we get

$$PAP^{-1}(Px) = \lambda PBP^{-1}(Px),$$

which establishes the first part. Conversely, premultiplying (21) by P^{-1}, we have $A(P^{-1}y) = \lambda B(P^{-1}y)$ which completes the proof.

Premultiplying (21) by J^+ and letting $C = J^+PAP^{-1}$ we obtain the equation $Cy = \lambda J^+Jy$, whose solutions we now investigate.

Theorem 12. Any solution $y = Px$ of equation (21), subject to the condition $B^SBx = x$ for some strong spectral inverse B^S, is an eigenvector of the matrix $C = J^+PAP^{-1}$, λ being the corresponding eigenvalue. Moreover, any eigenvector $y = Px$ of C is a solution of (21), λ being the relative eigenvalue, if $BB^S(Ax) = Ax$ holds for some strong spectral inverse B^S.

Proof. Premultiplying (21) by J^+ we have $J^+PAP^{-1}y = \lambda J^+Jy$. Using $B = P^{-1}JP$ and $B^S = P^{-1}J^+P$ we get

$$J^+PAP^{-1}y = \lambda PB^SB(P^{-1}y) = \lambda y$$

by the condition $B^SBx = x$. To establish the second part, premultiply $J^+PAP^{-1}y = \lambda y$ by J. Substituting for J and J^+ we obtain

$$PBB^SAP^{-1}y = \lambda PBP^{-1}y, \quad \text{or} \quad PBB^SAx = \lambda PBP^{-1}y.$$

Applying the condition in the theorem yields the desired result.

2.2. Incidence Matrices

An incidence matrix is a matrix that has only two nonzero entries which are 1 and -1 in each column of the matrix and has no zero rows. Two rows of an incidence matrix are said to be directly connected with each other if there is a column that has nonzero entries in both rows. Two rows, i and j, are indirectly connected if there is a sequence of rows, starting with the ith row and ending with the jth row, in which every two adjacent rows in the sequence is directly connected. Any two rows are said to be connected if they are directly or indirectly connected.

A connected component of an incidence matrix is a submatrix which consists of a set composed of rows, each pair of which is connected and none of which is connected with any other rows not in the set, and a set composed of all the columns that have nonzero entries in the rows in the set.

An incidence matrix is said to be a connected incidence matrix if it has only one connected component; otherwise it is said to be a separable incidence matrix.

By suitable row and column interchanges any separable incidence matrix can be expressed as a direct sum of the matrices of the connected components.

In the sequel we deal only with connected incidence matrices since the pseudoinverse of a separable incidence matrix can be obtained by adjoining a set of the pseudoinverses of its connected components using the fact that for permutation matrices P_1 and P_2, $(P_1AP_2)^+ = P_2^*A^+P_1^*$.

Theorem 13. Let A be an $m \times n$ connected incidence matrix, then

$$I - AA^+ = \frac{1}{m} E$$

where E is the matrix whose elements are all equal to 1.

Proof. Since $(I - AA^+)A = 0$, the kth column of A has 1 in the ith row and -1 in the jth row so the elements of $I - AA^+$ for columns i and j must be the same. Since A is connected, all columns of $I \quad AA^+$ are identical, and being hermitian, all rows of $I - AA^+$ are identical. Therefore the elements in $I - AA^+$ are all identical. Moreover, since $I - AA^+$ is idempotent, all its elements are equal to $1/m$.

Theorem 14. Let A be an $m \times n$ connected incidence matrix. A contains at least one set of $m - 1$ columns which form a basis for $R(A)$.

Proof. Let R_0 be an arbitrary row in A. Let R_1 be the set of all rows directly connected with R_0, and R_i the nonempty set of all rows not in $R_0 \cup R_1 \cup \cdots \cup R_{i-1}$ which are directly connected with at least one of the rows in R_{i-1}. Since A is connected and $R_i \cap R_j = 0$ for $i \neq j$, every row of A belongs to one and only one R_j, $1 \leq j \leq m - 1$. Select a column for every row in R_i that connects the row with any one of the rows in R_{i-1}, and let C_i be the set of such columns. Thus the number of columns in C_i equals the number of rows in R_i and $C_i \cap C_j = 0$ for $i \neq j$. Therefore $C = C_1 \cup C_2 \cup \cdots \cup C_j$ consists of $m - 1$ columns. Now every row is connected with R_0 by columns in C so that for any two rows in A, for example, the r_1th and r_jth rows, there is a sequence of columns in C, for instance, $c_1, c_2, \ldots, c_{j-1}$, where c_i, $i = 1, 2, \ldots, j - 1$, directly connects the r_ith and r_{i+1}th rows. Assume that in the sequence of rows $r_1, \ldots, r_j$ no two rows are identical, since in that case the sequence can be shortened by eliminating the intervening rows. This implies that no two columns are identical in the sequence of columns $c_1, \ldots, c_{j-1}$. If the kth column of A has a 1 in the r_1th row and -1 in the r_jth row, then this column is expressed as

$$a_k = \sum_{i=1}^{j-1} \pm c_i, \tag{22}$$

where the plus sign is taken if c_i has 1 in the r_ith row and the minus sign is taken if c_i has -1 in the r_ith row. Note that every row in R_i has only one column in $C_1 \cup C_2 \cup \cdots \cup C_i$, $i = 1, 2, \ldots, j$, which has a nonzero element in the row. Thus, if a linear combination of columns in C is equal to the zero vector, the coefficients in the linear combination for the columns in C_j must all be equal to zero. This implies that the coefficients for the columns in C_{k-1} must also be all equal to zero, which in turn implies that the coefficients for the columns in C_{k-2} must also be all equal to zero, and so on. It follows that C is linearly independent and therefore any column of A can be expressed uniquely as a linear combination of columns in C.

Some properties of the pseudoinverse of an incidence matrix are now derived.

First, define the directed graph of a connected incidence matrix as a graph whose vertices and arcs have a one-to-one correspondence with rows and columns, respectively, of the incidence matrix with each arc being directed from the ith vertex to the jth vertex if the corresponding column of the matrix has a -1 in the ith row and 1 in the jth row.

Let x be an n-component column vector that represents flows through n arcs in the graph, and Ax represents the amount of the net inflow (or outflow if negative) to be made at each one of the m vertices.

Since $AA^+ = I - 1/mE$, the ith column of A^+ represents the quantities that flow through the n arcs when one unit of inflow is made at the ith vertex and $1/m$ units of outflow are made at each of the m vertices. Moreover, from Theorem 14, any column of A can be uniquely expressed as the sign-adjusted sum of columns in a basis so that the corresponding row of A^+ is also uniquely expressible by the same sign-adjusted sum of the corresponding rows in the basis of A^+. (See problem 9.)

Thus the flow quantity in the jth arc is equal to the sign-adjusted sum of the flow quantities in a sequence of arcs whose corresponding columns are in the basis and connect the same two vertices as the jth arc does. Hence it follows that the sign-adjusted sum of the flow quantities in any sequence of arcs which connect a pair of vertices is identical for any given pair of vertices.

It will now be shown that the element in the jth row and the ith column of A^+ is equal to the quantity that flows through the jth arc in the direction of the arc if the flows in the graph are made in such a way that one unit of inflow is made at the ith vertex and $1/m$ units of outflow are made at each of the m vertices, and for every pair of vertices the sign-adjusted sum of the flow quantities in a sequence of arcs which connect the two vertices is identical for any such sequences. Furthermore, these two conditions uniquely determine A^+ for any given directed graph in which the correspondence between rows of A^+ and arcs of the graph and the correspondence between columns of A^+ and vertices of the graph are fixed.

To establish these results, let A be partitioned as $[U, V]$ where U is an $m \times (m-1)$ matrix whose columns form a basis and V has columns not in the basis. Let D be the matrix such that $V = UD$. Suppose that two $n \times m$ matrices X and Y satisfy the above two conditions. Then, by the first condition, $A(X - Y) = 0$. Let the matrix $X - Y$ be partitioned in

$$\begin{pmatrix} W \\ Z \end{pmatrix}$$

where W is an $(m-1) \times m$ matrix which corresponds to basis arcs and Z is an $(n - m + 1) \times m$ matrix which corresponds to nonbasis arcs. It follows from the second condition that $Z = D^*W$. Hence $A(X - Y) = UW$ $+ UDD^*W = U(I + DD^*)W = 0$. Since the columns of U are linearly independent and $I + DD^*$ is nonsingular, it follows that $W = 0$. Hence the matrix that satisfies the two conditions for a given graph is unique.

EXERCISES

1. Simplify the sequential algorithm for obtaining $\hat{x}_n$ when the observations are scalars.
2. Consider the model $QAx = Qb + Qe$ where Q is nonsingular. Derive the sequential weighted least squares algorithm with weighting matrix Q^*Q.
3. Assuming that Qe in problem 2 is a random vector with mean 0 and covariance matrix I, obtain an expression for the covariance matrix $C(\hat{x}_n, \hat{x}_n)$ of the estimator $\hat{x}_n$. Derive an expression for $C(\hat{x}_n, \hat{x}_n)$ in terms of $C(\hat{x}_{n-1}, \hat{x}_{n-1})$ and the nth observation.
4. Consider the linear model $y = Hx + e$, where $E(e) = 0$, $E(ee^*) = V$, a positive definite hermitian matrix and H is $p \times n$. Prove:
   (a) If $r(H) = n$, the minimum variance linear unbiased estimate for x is

   $$\hat{x} = (H^*V^{-1}H)^{-1}H^*V^{-1}y$$

   with covariance matrix

   $$R_{\hat{x}} = (H^TV^{-1}H)^{-1}.$$

   (b) If $r(H) = r \leq \min(p, n)$, show that

   $$\hat{x} = (H^*V^{-1}H)^+H^*V^{-1}y$$

   is the best linear estimate of x, that is, $\hat{x}$ is a best linear estimate if $\|E(x) - x\|$ is minimum and $\hat{x}$ has minimum variance with respect to the range space of H^*.
   (c) Show that if $r = p$ in (b), then $\hat{x}$ becomes the least squares estimate of x.
5. Prove Theorem 3 with C^+ replaced by an arbitrary generalized inverse of C.
6. Show that any solution x of (19) contained in $R(B)^*$ is an eigenvector of B^+A, λ being the relative eigenvalue. Conversely, an eigenvector x of B^+A is a solution of (19), λ being the corresponding eigenvalue if $Ax \in R(B)$.
7. Construct the directed graph of the following incidence matrix:

$$A = \begin{bmatrix} 1 & 0 & 0 & -1 \\ -1 & 1 & 0 & 1 \\ 0 & -1 & -1 & 0 \end{bmatrix}.$$

Calculate A^+ and completely label the flows through the arcs of the directed graph.
8. Prove that since any column of the connected incidence matrix A can be uniquely expressed as the sign-adjusted sum of columns in a basis, the corresponding row of A^+ is also uniquely expressible by the same sign-adjusted sum of the corresponding rows in the basis of A^+. *Hint:* Use $Ax = 0$ if and only if $x^*A^+ = 0$.

APPENDIX 1
COMPUTING TECHNIQUES

In recent years many techniques for obtaining the pseudoinverse of a matrix have appeared in the literature. It is the purpose of this appendix briefly to acquaint an individual with two of these algorithms without going into the vexing problems of rounding error, word length, double and triple precision arithmetic, and so on. The interested reader is encouraged to consult the references for additional information of this nature.

1. A DIRECT METHOD

Hestenes [1] obtained a scheme for inverting matrices by a process called biorthogonalization. This process can be extended and modified to all rectangular matrices of any rank.

The concept of biorthogonality can be expressed in matrix form as follows. The vectors (in an n-dimensional space) $u_1, u_2, \ldots, u_n$ can be considered to be the column vectors of a matrix U and $v_1, \ldots, v_n$, the row vectors of a matrix V. The set is biorthogonal if $VU = I$. If $n = m$, then V is the inverse of U.

Hestenes poses and solves the problem: given two sets $u_1, \ldots, u_n$ and $v_1, \ldots, v_n$ of n vectors in an m-dimensional space $(m \geq n)$, obtain a biorthogonal system by modifying the vs. The solution is arrived at by letting $v_1^{(0)}, \ldots, v_n^{(0)}$ be the initial choice for the vs. These vectors are modified successively in n steps. After n steps the vectors $v_1^{(n)}, \ldots, v_n^{(n)}$ will be a solution to the problem.

In the kth step the vectors $v_i^{(k-1)}$ are transformed into a new set $v_i^{(k)}$ by the following computations:

$$c_{kk} = \langle v_k^{(k-1)}, u_k \rangle \quad \text{where} \quad \langle a, b \rangle = a^T b,$$
$$c_k = c_{kk}^{-1},$$
$$c_{jk} = \langle v_j^{(k-1)}, u_k \rangle \quad j \neq k,$$
$$v^{(k)} = c_k v^{(k-1)}$$

where $c_{ij}^{(k)} = \delta_{ij}(j \neq k)$, $c_{kk}^{(k)} = -c_k$, $c_{ik}^{(k)} = c_{ik} c_k$ for $i \neq k$.

The only difficulty is to insure that $c_{kk} \neq 0$. This will not arise if $V^{(0)}$ is taken to be U^*.

To compute the pseudoinverse of a matrix, the method is modified by adding rows to the original matrix which are orthogonal and which raise the rank of the matrix to its column dimension. Then the method is applied to the resulting matrix. The pseudoinverse of the original matrix is obtained by deleting the last added columns of $V^{(n)}$.

Example. Let

$$A = \begin{pmatrix} 1 & 0 & 1 & 1 \\ 0 & 1 & -1 & 0 \\ 1 & 1 & 0 & 1 \end{pmatrix}.$$

Add two rows to A which are orthogonal to all other rows.

$$U = \begin{pmatrix} 1 & 0 & 1 & 1 \\ 0 & 1 & -1 & 0 \\ 1 & 1 & 0 & 1 \\ \hdotsfor{4} \\ 1 & 0 & 0 & -1 \\ 0 & 1 & 1 & -1 \end{pmatrix}$$

and let

$$V^{(0)} = U^* = \begin{pmatrix} 1 & 0 & 1 & 1 & 0 \\ 0 & 1 & 1 & 0 & 1 \\ 1 & -1 & 0 & 0 & 1 \\ 1 & 0 & 1 & -1 & -1 \end{pmatrix}.$$

Then

$$c_{11} = \langle v_1^{(0)}, u_1 \rangle = (1, 0, 1, 1, 0)(1, 0, 1, 1, 0)^T = 3$$

so that $c_1 = \tfrac{1}{3}$.

$$c_{21} = \langle v_2^{(0)}, u_1 \rangle = (0, 1, 1, 0, 1)(1, 0, 1, 1, 0)^T = 1,$$
$$c_{21}^{(1)} = -c_{21} c_1 = -\tfrac{1}{3},$$

$$c_{31} = \langle v_3^{(0)}, u_1 \rangle = (1, -1, 0, 0, 1)(1, 0, 1, 1, 0)^T = 1,$$

$$c_{31}^{(1)} = -c_{31}c_1 = -\tfrac{1}{3},$$

$$c_{41} = \langle v_4^{(0)}, u_1 \rangle = (1, 0, 1, -1)(1, 0, 1, 1, 0)^T = 1,$$

$$c_{41}^{(0)} = -c_{41}c_1 = -\tfrac{1}{3}.$$

Hence

$$c^{(1)} = \begin{pmatrix} \tfrac{1}{3} & 0 & 0 & 0 \\ -\tfrac{1}{3} & 1 & 0 & 0 \\ -\tfrac{1}{3} & 0 & 1 & 0 \\ -\tfrac{1}{3} & 0 & 0 & 1 \end{pmatrix}$$

and

$$V_1^{(1)} = \begin{pmatrix} \tfrac{1}{3} & 0 & 0 & 0 \\ -\tfrac{1}{3} & 1 & 0 & 0 \\ -\tfrac{1}{3} & 0 & 1 & 0 \\ -\tfrac{1}{3} & 0 & 0 & 1 \end{pmatrix} \begin{pmatrix} 1 & 0 & 1 & 1 & 0 \\ 0 & 1 & 1 & 0 & 1 \\ 1 & -1 & 0 & 0 & 1 \\ 1 & 0 & 1 & -1 & -1 \end{pmatrix}$$

$$= \begin{pmatrix} \tfrac{1}{3} & 0 & \tfrac{1}{3} & \tfrac{1}{3} & 0 \\ -\tfrac{1}{3} & 1 & \tfrac{2}{3} & -\tfrac{1}{3} & 1 \\ \tfrac{2}{3} & -1 & -\tfrac{1}{3} & -\tfrac{1}{3} & 1 \\ \tfrac{2}{3} & 0 & \tfrac{2}{3} & -\tfrac{4}{3} & -1 \end{pmatrix}.$$

Now

$$c_{22} = \langle v_2^{(1)}, u_2 \rangle = (-\tfrac{1}{3}, 1, \tfrac{2}{3}, -\tfrac{1}{3}, 1)(0, 1, 1, 0, 1)^T = \tfrac{8}{3}, \ c_{22}^{(2)} = \tfrac{3}{8},$$

$$c_{12} = \langle v_1^{(1)}, u_2 \rangle = (\tfrac{1}{3}, 0, \tfrac{1}{3}, \tfrac{1}{3}, 0)(0, 1, 1, 0, 1)^T = \tfrac{1}{3}, \ c_{12}^{(2)} = -\tfrac{1}{8},$$

$$c_{32} = \langle v_3^{(1)}, u_2 \rangle = (\tfrac{2}{3}, -1, -\tfrac{1}{3}, -\tfrac{1}{3}, 1)(0, 1, 1, 0, 1)^T = -\tfrac{1}{3}, \ c_{32}^{(2)} = \tfrac{1}{8},$$

$$c_{42} = \langle v_4^{(1)}, u_2 \rangle = (\tfrac{2}{3}, 0, \tfrac{2}{3}, -\tfrac{4}{3}, -1)(0, 1, 1, 0, 1)^T = -\tfrac{1}{3}, \ c_{42}^{(2)} = \tfrac{1}{8}.$$

Hence

$$V^{(2)} = C^{(2)}V^{(1)} = \begin{pmatrix} 1 & -\tfrac{1}{8} & 0 & 0 \\ 0 & \tfrac{3}{8} & 0 & 0 \\ 0 & \tfrac{1}{8} & 1 & 0 \\ 0 & \tfrac{1}{8} & 0 & 1 \end{pmatrix} \begin{pmatrix} \tfrac{1}{3} & 0 & \tfrac{1}{3} & \tfrac{1}{3} & 0 \\ -\tfrac{1}{3} & 1 & \tfrac{2}{3} & -\tfrac{1}{3} & 1 \\ \tfrac{2}{3} & -1 & -\tfrac{1}{3} & -\tfrac{1}{3} & 1 \\ \tfrac{2}{3} & 0 & \tfrac{2}{3} & -\tfrac{4}{3} & -1 \end{pmatrix}$$

$$= \begin{pmatrix} \tfrac{3}{8} & -\tfrac{1}{8} & \tfrac{2}{8} & \tfrac{3}{8} & -\tfrac{1}{8} \\ -\tfrac{1}{8} & \tfrac{3}{8} & \tfrac{2}{8} & -\tfrac{1}{8} & \tfrac{3}{8} \\ \tfrac{5}{8} & -\tfrac{7}{8} & -\tfrac{2}{8} & -\tfrac{3}{8} & \tfrac{9}{8} \\ \tfrac{5}{8} & \tfrac{1}{8} & \tfrac{6}{8} & -\tfrac{11}{8} & -\tfrac{7}{8} \end{pmatrix}.$$

Likewise

$$V^{(3)} = \begin{pmatrix} \frac{6}{21} & 0 & \frac{6}{21} & \frac{9}{21} & -\frac{6}{21} \\ -\frac{2}{21} & \frac{7}{21} & \frac{5}{21} & -\frac{3}{21} & \frac{9}{21} \\ \frac{5}{21} & -\frac{7}{21} & -\frac{2}{21} & -\frac{3}{21} & \frac{9}{21} \\ \frac{15}{21} & 0 & \frac{15}{21} & -\frac{30}{21} & -\frac{15}{21} \end{pmatrix}$$

and

$$V^{(4)} = \begin{pmatrix} \frac{1}{5} & 0 & \frac{1}{5} & \frac{3}{5} & -\frac{1}{5} \\ -\frac{1}{15} & \frac{1}{3} & \frac{4}{15} & -\frac{1}{5} & \frac{2}{5} \\ \frac{4}{15} & -\frac{1}{3} & -\frac{1}{15} & -\frac{1}{5} & \frac{2}{5} \\ \frac{1}{5} & 0 & \frac{1}{5} & -\frac{2}{5} & -\frac{1}{5} \end{pmatrix}.$$

Hence the pseudoinverse of A is obtained by deleting the last two columns of $V^{(4)}$.

$$A^+ = \begin{pmatrix} \frac{1}{5} & 0 & \frac{1}{5} \\ -\frac{1}{15} & \frac{1}{3} & \frac{4}{15} \\ \frac{4}{15} & -\frac{1}{3} & -\frac{1}{15} \\ \frac{1}{5} & 0 & \frac{1}{5} \end{pmatrix}.$$

2. AN ITERATIVE METHOD

Greville [5] presented a concise recursive algorithm for computing the pseudoinverse of a matrix. The algorithm is as follows:

Let a_k denote the kth column of a given matrix A, and let A_k denote the matrix consisting of the first k columns. Consider A_k in the partitioned form (A_{k-1}, a_k).

Compute $d_k = A_{k-1}^+ a_k$ and $c_k = a_k - A_{k-1} d_k$. If $c_k \neq 0$, let $b_k = c_k^+$. If $c_k = 0$, compute $b_k = (1 + d_k^T d_k)^{-1} d_k^T A_{k-1}^+$. Then

$$A_k^+ = \begin{pmatrix} A_{k-1}^+ - d_k b_k \\ b_k \end{pmatrix}.$$

To initiate the process, take $A_1^+ = 0$ if a_1 is a zero vector; otherwise $A_1^+ = (a_1^T a_1)^{-1} a_1^T$.

This algorithm is very easy to follow and compute. It requires only one decision in each cycle. It requires no inverse of a matrix.

Example. Let

$$A = \begin{pmatrix} 1 & 0 & 1 & 1 \\ 0 & 1 & -1 & 0 \\ 1 & 1 & 0 & 1 \end{pmatrix},$$

$$A_1 = \begin{pmatrix} 1 \\ 0 \\ 1 \end{pmatrix}, \quad a_1 = \begin{pmatrix} 1 \\ 0 \\ 1 \end{pmatrix},$$

then $A_1^+ = (\tfrac{1}{2}, 0, \tfrac{1}{2})$.

$$d_2 = A_1^+ a_2 = (\tfrac{1}{2}, 0, \tfrac{1}{2})\begin{pmatrix} 0 \\ 1 \\ 1 \end{pmatrix} = (\tfrac{1}{2}),$$

$$c_2 = a_2 - A_1 d_2 = \begin{pmatrix} -\tfrac{1}{2} \\ 1 \\ \tfrac{1}{2} \end{pmatrix} \neq 0.$$

Hence $b_2 = c_2^+ = (-\tfrac{1}{2}, \tfrac{2}{3}, \tfrac{1}{3})$ and

$$A_2^+ = \begin{pmatrix} A_1^+ - d_2 b_2 \\ b_2 \end{pmatrix} = \begin{pmatrix} \tfrac{2}{3} & -\tfrac{1}{3} & \tfrac{1}{3} \\ -\tfrac{1}{3} & \tfrac{2}{3} & \tfrac{1}{3} \end{pmatrix}.$$

Now

$$d_3 = A_2^+ a_3 = \begin{pmatrix} 1 \\ -1 \end{pmatrix},$$

$$c_3 = a_3 - A_2 d_3 = \begin{pmatrix} 1 \\ -1 \\ 0 \end{pmatrix} - \begin{pmatrix} 1 & 0 \\ 0 & 1 \\ 1 & 1 \end{pmatrix}\begin{pmatrix} 1 \\ -1 \end{pmatrix} = 0.$$

Hence

$$b_3 = (1 + d_3^T d_3)^{-1} d_3^T A_2^+$$

$$= (1 + 2)^{-1}(1, -1)\begin{pmatrix} \tfrac{2}{3} & -\tfrac{1}{3} & \tfrac{1}{3} \\ -\tfrac{1}{3} & \tfrac{2}{3} & \tfrac{1}{3} \end{pmatrix}$$

$$= (\tfrac{1}{3}, -\tfrac{1}{3}, 0).$$

Thus

$$A_3^+ = \begin{pmatrix} A_2^+ - d_3 b_3 \\ b_3 \end{pmatrix} = \begin{pmatrix} \tfrac{1}{3} & 0 & \tfrac{1}{3} \\ 0 & \tfrac{1}{3} & \tfrac{1}{3} \\ \tfrac{1}{3} & -\tfrac{1}{3} & 0 \end{pmatrix}.$$

Now

$$d_4 = \begin{pmatrix} \tfrac{2}{3} \\ \tfrac{1}{3} \\ \tfrac{1}{3} \end{pmatrix}$$

and $c_4 = 0$ so that $b_4 = (\tfrac{1}{3}, 0, \tfrac{1}{3})$ and

$$A_4^+ = \begin{pmatrix} A_3^+ - d_4 b_4 \\ b_4 \end{pmatrix} = \frac{1}{15}\begin{pmatrix} 3 & 0 & 3 \\ -1 & 5 & 4 \\ 4 & -5 & -1 \\ 3 & 0 & 3 \end{pmatrix}.$$

APPENDIX 2
NOTES AND COMMENTS

1. CHAPTER 1

W. T. Reid [5] claims that the introduction of the general reciprocal or generalized inverse of an operator was not in the algebraic case considered by E. H. Moore, but in the setting of integral and differential operators. Reid in his interesting paper develops the history of the concept of generalized inverses in classical analysis. Generally it is accepted that E. H. Moore gave in 1920 a "useful extension of the classical notion of the reciprocal of a non-singular square matrix." In 1935 Moore discussed the concept at some length in his *General Analysis*. Parts of his work have been interpreted in modern notation by Ben-Israel and Charnes [2] and by Greville [1, 2]. It is interesting to note that in 1936 Von Neumann [2] gave an algebraic basis and extension in his studies on regular rings.

Unaware of Moore's results, Bjerhammar [1, 2] and Penrose [1, 2] gave independent developments of the pseudoinverse. Bjerhammar gave the general solution of $Ax = b$, when solvable, as $x = A^+b + (I - A^+A)y$ where y is arbitrary up to dimensional compatibility. The least square character of the solution was used by Bjerhammar in geodetic applications, adjusting observations which gave rise to singular matrices.

Penrose [1] defined the pseudoinverse as the unique solution of the equations (1) through (4) of Chapter 1.

The definitions (1–4) establish a hierarchy of generalized inverses and were apparently first collected by Rohde [1, 3]. Theorem 1 was first established by

Penrose [1]. Theorems 2 to 4 appear in a paper by Urguhart [1], and while Theorem 5 is due to Penrose, we present Greville's [10] version.

The concept of a weighted pseudoinverse was introduced by Chipman, and studied further by Miecler [1].

Lemma 1 is due to Penrose [1]. Lemmas 2 and 3 and Theorem 6 appear in a paper by Morris and Odell [1]. Part of the properties listed in Section 3 are those reported by Price.

2. CHAPTER 2

The results in this chapter, with few exceptions, are those obtained by Cline [3, 4]. It is worthy of comment that there exist a few among us who attack the most difficult questions in terms of tedious and complicated manipulations. Cline has taken on these types of questions and gained important results in the form of identities which have made various theoretical results easy to obtain.

Theorems 1 and 3 through 7 are due to Cline [2–4], while Theorem 2 is a result reported by Greville [10]. Greville's work is written in a succinct and easily understood manner. The young graduate student should read Greville's works early in his studies of generalized inverses.

3. CHAPTER 3

Motivated by Greville's papers [4–5] on applying the generalized inverse to solutions of linear equations involving probabilities, Odell in 1963 discovered a need for what Greville later called a spectral inverse in [11, 12]. Scroggs and Odell were the first to consider defining a unique spectral inverse. An error in the paper was later corrected by Boullion and Odell [3]. It was Greville, however, who gave a reasonable formulation of a spectral inverse. It should be noted that Drazin defined a generalized inverse, sometimes called the Drazin inverse, which has spectral properties. Erdelyi [4] defined a solution of (17) to (20), when it exists, as a quasi-commuting inverse. This inverse was shown to be equivalent to the spectral inverse by Greville [12].

Rohde gave Definitions 1 and 2 in his dissertation [1] and paper [3]. Theorem 1 is due to Erdelyi [5]. Theorem 2 is Drazin's result. Theorem 3 can be found in the paper by Erdelyi [4].

Greville [12] first formulated Definitions 3 through 6 and from these he proved Theorems 4 through 7. The concept of strong spectral inverse is Greville's generalization which followed naturally from Scroggs and Odell.

4. CHAPTER 4

The number of papers that are appearing in the journals is increasing rapidly, and it is difficult to select special topics which are most important. New results will overshadow older ones. However, we chose those topics that appeared to be receiving the most attention by researchers prior to 1969.

Range hermitian matrices were first studied by Schwerdtfeger and by Pearl [1–3], both of whom called them EPr matrices. The term range-hermitian was suggested by Greville. Theorems 1 and 2 are due to Hearon [1].

Section 2 of this chapter is principally based on the work of Erdelyi [2]. Section 3 is based on a paper by Hearon and Evans [1].

5. CHAPTER 5

Perhaps the major general application of the theory of generalized inverses of matrices is in solving the matrix equation $AX = B$ for the matrix X. This is the task that Bjerhammar [1] wished to perform, and the usefulness that Penrose noted [2]. Section 1 contains these early results of Penrose.

Section 2 represents the extension of Penrose's ideas to more general norms. Odell conceived the idea of solutions that minimized various norms. Meicler [1] studied these and wrote a dissertation on the topic. It was not until Newman developed the notation used here that the theory was presented in a desirable fashion. The results in this section were reported previously in the paper by Newman and Odell.

Section 3 reports some results obtained by Morris and Odell [2]. These results were developed in an effort to produce sequential techniques for computing the pseudoinverse of a matrix.

6. CHAPTER 6

One is indeed brave to select special topics for applications. The sequential least squares parameter estimation scheme was obtained by Rainbolt. Other results related to this can be found in the papers by Albert and Sittler, Rohde and Harvey, and Chipman.

Theorem 2 is due to C. R. Rao [4]. Theorem 3 is due to Marsaglia. Interesting related topics can be found in the papers by Searle [1], Price, Frame, Gately, John, Goldman and Zelen, Harris and Helvig [1, 2], Lewis and Odell [1, 2], Mitra [1, 2], and C. R. Rao [1–4].

The material in Section 1.3 was recently obtained by T. Watkins, T. O. Lewis, and T. L. Boullion.

Section 1.4 is based on a paper by Decell and Odell [2]. The computation of the probability vector is discussed in the paper by Odell and Decell.

Section 1.5 is based on part of the paper by Scroggs and Odell.

Section 2.1 is based on a paper by Erdelyi [5] and Section 2.2 is based on a paper by Ijiri.

Many other applications have appeared in the literature. A generalized inverse was utilized by Meyer [3] to obtain solutions to the system of equations $AX = A$, $YA = A$, and by Jameson to solve the system $AX + XB = C$. Pyle [1, 2] has applied a generalized inverse to linear programming problems. Nelson has obtained significant results on nonlinear programming using the pseudoinverse. Kirby has employed it in chance constrained programming problems. Ben Israel [10] has obtained solutions to 2-person 0-sum games using the pseudoinverse. Bott and Duffin used a generalized inverse in their paper on the algebra of networks. Graybill [2] has employed generalized inverses throughout his book on matrix theory with applications in statistics.

REFERENCES

Afriat, S. N.
"Orthogonal and oblique projectors and the characteristics of pairs of vector spaces,"
Proc. Cambridge Philos. Soc., LIII (1957), 800–816.

Aitken, A. C.
"On least squares and linear combinations of observations," *Proc. Roy. Soc. Edinburgh Sect. A.*, LV (1934), 42–48.

Albert, A.
[1] "An introduction and beginner's guide to matrix pseudo inverses," *ARCON— Advanced Research Consultants*, Lexington, Mass., July 1964.

[2] "Real time computation of constrained least squares estimators," *ARCON— Advanced Research Consultants*, Lexington, Mass., June 1965.

Albert, A., and R. W. Sittler
"A method for computing least squares estimators that keep up with the data," *J. Soc. Indust. Appl. Math. Ser. A Control 3* (1965), 384–417.

Altman, M.
"On the approximate solution of linear algebraic equations," *Bull. Acad. Polon. Sci. Ser. Sci. Math. Astronom. Phys., Cl. III*, V (1957), 365–370.

Anderson, C. L.
"A geometric theory of pseudoinverses and some applications in statistics," Master's thesis, Southern Methodist University, Dallas, Texas, 1967.

Anderson, W. N., Jr., and R. J. Duffin
"Series and parallel addition of matrices," *J. Math. Anal. Appl.*, XXVI, no. 3 (June 1969), 576–594.

Arghiriade, E.
[1] "Sur les matrices qui sont permutables avec leur inverse généralisée," *Atti Accad. Naz. Lincei Rend. Cl. Sci. Fis. Mat. Natur.*, XXV, no. 8 (1963), 244–251.

81

[2] "Remarques sur l'inverse généralisée d'un produit de matrices," *Atti Accad. Naz. Lincei Rend. Cl. Sci. Fis. Mat. Natur.*, XLII, no. 5 (1967), 621–625.

Arghiriade, E., and A. Dragomir
"Une nouvelle définition de l'inverse généralisée d'une matrice," *Atti Accad. Naz. Lincei Rend. Cl. Sci Fis. Mat. Natur.*, XXXV, no. 8 (1963), 158–163.

Autonne, L.
"Sur les matrices hypohermitiennes et sur les matrices unitaries," *Ann. Univ. Lyon Sect. A*, XXXVIII (1917), 1–77.

Azumaya, G.
"Strongly π-regular rings," *J. Fac. Sci. Hokkaido Univ. Ser. 1*, XIII (1954), 34–39.

Balakrishnan, A. V.
"An operator theoretic formulation of a class of control problems and a steepest descent method of solution," *J. Soc. Indust. Appl. Math. Ser. A Control 1* (1963), 109–127.

Banerjee, K. S.
"Singularity in Hotelling's weighing designs and a generalized inverse," *Ann. Math. Statist.*, XXXVII, no. 4 (1964), 1021–1032.

Barton, C. P.
"Pseudo inverses of rectangular matrices," Master's thesis, University of Texas, Austin, Texas, 1966.

Baskett, T. S., and I. J. Katz
"Theorems on products of EPr matrices," *J. Lin. Alg. Appl.*, in press.

Bauer, F. L.
"Elimination with weighted row combinations for solving linear equations and least squares problems," *Numer. Math.*, VII (1965), 338–352.

Bellman, R.
Introduction to Matrix Analysis. New York: McGraw-Hill, 1960.

Ben-Israel, A.
[1] "On direct sum decompositions of Hestenes algebras," *Israel J. Math.*, II, no. 1 (1964), 50–54.

[2] "An iterative method for computing the generalized inverse of an arbitrary matrix," *Math. Comp.*, XIX, no. 91 (1965), 452–455.

[3] "A modified Newton-Raphson method for the solution of systems of equations," *Israel J. Math.*, III (1965), 94–98.

[4] "A note on an iterative method for generalized inversion of matrices," *Math. Comp.*, XX, no. 95 (1966), 439–440.

[5] "A Newton-Raphson method for the solution of systems of equations," *J. Math. Anal. Appl.*, XV (1966), 243–252.

[6] "A note on the Cayley transform," *Notices Amer. Math. Soc.*, XIII (1966), 599.

[7] "On error bounds for generalized inverses." *J. Soc. Indust. Appl. Math. Numer. Anal.*, III, no. 4 (1966), 585–592.

[8] "On the geometry of subspaces in Euclidean n-spaces," *J. Soc. Indust. Appl. Math.*, XV, no. 5 (September 1967, 1189–1198.

[9] "On iterative methods for solving nonlinear least squares problems over convex sets," *Israel J. Math.*, V (1967), 211–224.

[10] "On optimal solutions of 2-person 0-sum games." Evanston, Ill.: Systems Research Memo no. 195, Northwestern University, March 1968.

[11] "An application of the Newton-Raphson method to the special eigenvalue problem," unpublished manuscript, May 1966.

[12] "A note on partitioned matrices and equations," *SIAM Rev*, XI (April 1969), 247–250.

[13] "Linear equations and inequalities on finite dimensional, real or complex, vector spaces: A unified theory," *J. Math. Anal. Appl.*, *XXVII*, no. 2 (August 1969), 367–389.

Ben-Israel, A., and A. Charnes
[1] "An explicit solution of a special class of linear programming problems." Evanston, Ill.: Systems Research Memo. no. 191, Northwestern University, December 1967.

[2] "Contributions to the theory of generalized inverses," *J. Soc. Indust. Appl. Math.*, XI (September 1963), 667–699.

[3] "Generalized inverses and the Bott-Duffin network analysis," *J. Math. Anal. Appl.*, VII (1963), 428–435. Erratum: *ibid.*

[4] "On the intersection of cones and subspaces," *Bull. Amer. Math. Soc.* (May 1968).

[5] "Projection properties and the Neumann-Euler expansion for the Moore-Penrose inverse of an arbitrary matrix." Evanston, Ill.: ONR Research Memo. No. 40, Northwestern University, The Technological Institute, 1961.

Ben-Israel, A., and D. Cohen
"On iterative computation of generalized inverses and associated projections," *J. Soc. Indust. Appl. Math. Numer. Anal.*, III (1966), 410–419.

Ben-Israel, A., and Y. Ijiri
"A report on the machine calculation of the generalized inverse of an arbitrary matrix." Pittsburgh, Pa.: ONR Research Memo. no. 110, Carnegie Institute of Technology March 1963.

Ben-Israel, A., and S. J. Wersan
"An elimination method for computing the generalized inverse of an arbitrary complex matrix," *J. Asso. Comput. Mach.*, X (1963), 532–537.

Berge, Claude
Theorie des Graphes et ses Applications. Paris: Dunod, 1958. Translation: *The Theory of Graphs and Its Applications*. New York: Wiley, 1962.

Beutler, F. J.
[1] "The operator theory of the pseudo-inverse, I. Bounded operators," *J. Math. Anal. Appl.*, X (1965), 451–470.

[2] "The operator theory of the pseudoinverse, II. Unbounded operators with arbitrary range," *J. Math. Anal. Appl.*, X (1965), 471–493.

Bjerhammar, A.
[1] "Rectangular reciprocal matrices with special reference to geodetic calculations *Bull. Géodésique* (1951), 188–220.

[2] "Application of the calculus of matrices to the method of least squares, with special reference to geodetic calculations," *Kung. Tekn. Hogsk. Handl. Stockholm*, XLIX (1951), 1–68.

[3] "A generalized matrix algebra," *Kung. Tekn. Hogsk. Handl. Stockholm*, CXXIV (1958), 1–32.

Blattner, J. W.
"Border matrices," *J. Soc. Indust. Appl. Math.*, X (1962), 528–536.

Bliss, G. A.
Lectures on Calculus of Variation. Chicago: University of Chicago Press, 1946.

Boot, J. C. G.
[1] "The computation of the generalized inverse of singular or rectangular matrices," *Amer. Math. Monthly*, LXX (1963), 302–303.

[2] "Projection matrices and the generalized inverse." Buffalo, N. Y.: State University of New York at Buffalo, March 1965.

Boros, Emil
"On the generalized inverse of an EPr matrix," *An. Univ. Timisoara Ser. Sti. Mat.-Fiz.*, no. 2 (1964), 33–38.

Boros, E., and I. Sturz
"On quasi-inverse matrices," *An. Univ. Timisoara Ser. Sti. Mat.-Fiz.*, no. 1 (1963), 59–66.

Bott, R., and R. J. Duffin
"On the algebra of networks," *Trans. Amer. Math. Soc.*, LXXIV (1953), 99–109.

Boullion, T. L.
"Contributions to the theory of pseudoinverses," Doctoral dissertation, University of Texas, Austin, Texas, 1966.

Boullion, T. L., and P. L. Odell
[1] "A generalization of the Wielandt inequality," *Texas Journal of Science*, XX, no. 3 (February 1969), 255–260.

[2] "An introduction to the theory of generalized matrix invertibility." Austin, Texas: Texas Center for Research, 1966.

[3] "A note on the Scroggs-Odell pseudoinverse," *J. Soc. Indust. Appl. Math.*, XVII (January 1969), 7–10.

[4] *Theory and Application of Generalized Inverses of Matrices*, Symposium Proceedings Texas Tech University Mathematics Series no. 4, Lubbock, Texas.

Bounitzky, E.
"Sur la fonction de Green des equations differentielles lineaires ordinaires," *J. Math. Pures Appl.* (6), V (1909), 65–125.

Bradley, J. S.
[1] "Adjoint quasi-differential operators of Euler type," *Pacific J. Math.*, XVI (1966), 213–237.

[2] "Generalized Green's matrices for compatible differential systems," *Michigan Math. J.*, XIII (1966), 97–108.

Brand, L.
"The solution of linear algebraic equations," *Math. Gaz.*, XLVI (1962), 203–207.

Den Broeder, G. G., Jr., and A. Charnes
"Contributions to the theory of generalized inverses for matrices." Lafayette, Ind.: Purdue University, 1957. (Reprinted as *ONR Res. Memo.* no. 39, Northwestern University, Evanston, Ill., 1962.)

Bryson, A. E., and D. E. Johanson
"Linear filtering for varying systems using measurements containing colored noise," unpublished report.

References   85

Charnes, A., and W. W. Cooper
Management Models and Industrial Applications of Linear Programming. New York: Wiley, 1961.

Charnes, A., W. W. Cooper, and G. L. Thompson
[1] "Constrained generalized medians and hypermedians as deterministic equivalents for two-stage linear programs under uncertainty," *Management Sci.*, XII (1965), 83–112.

(2) "Constrained generalized median and linear programming under uncertainty." Evanston, Ill. and Pittsburgh, Pa.: ONR Research Memo. no. 41, Northwestern University, The Tehnological Institute, and Carnegie Institute of Technology, 1962.

Charnes, A., and M. Kirby
[1] "A linear programming application of a left inverse of a basis matrix." Evanston, Ill.: ONR Research Memo. no. 91, Northwestern University, November 1963.

[2] "Modular design, generalized inverses, and convex programming," *Operations Res.*, XIII (1965), 836–847.

Chernoff, Herman
"Locally optimal designs for estimating parameters," *Ann. Math. Statist.*, XXIV (1934), 586–602.

Chipman, J. S.
"On least squares with insufficient observations," *J. Amer. Statist. Assoc.*, LIV (1964), 1078–1111.

Chipman, J. S., and M. M. Rao
[1] "Projections, generalized inverses, and quadratic forms," *J. Math. Anal. Appl.*, IX (1964), 1–11.

[2] "On the treatment of linear restrictions in regression analysis," *Econometrica*, XXXII (1964), 198–209.

Cho, C. Y.
"Talks on generalized inverses and solutions of large, approximately singular linear systems." Madison, Wis.: MRC Technical Summary Report no. 644, University of Wisconsin, Mathematics Research Center, U. S. Army, April 1966. (Omitted: R. E. Cline, T. N. E. Greville, B. Noble, L. D. Pyle, J. B. Rosen, and D. V. Steward.)

Clifford, A. H.
"Semigroups admitting relative inverses," *Ann. of Math.*, XLII (1941), 1037–1049.

Clifford, A. H., and G. B. Preston
The Algebraic Theory of Semigroups, I. Providence, R. I.: Mathematical Suverys, no. 7, American Math. Soc., 1961.

Cline, R. E.
[1] "Representations for the generalized inverse of matrices with applications in linear programming," Doctoral dissertation, Purdue University, Lafayette, Ind., 1963.

[2] "Note on the generalized inverse of the product of matrices," *SIAM Rev.*, VI (1964), 57–58.

[3] "Representations for the generalized inverse of a partitioned matrix," *J. Soc. Indust. Appl. Math.*, XII (1964), 588–600.

[4] "Representations for the generalized inverse of sums of matrices," *J. Soc. Indust. Appl. Math. Numer. Anal.*, II (1965), 99–114.

[5] "An application of representations for the generalized inverse of a matrix." Madison, Wis.: MRC Technical Summary Report no. 592, University of Wisconsin, Math. Research Center, U. S. Army, September 1965.

[6] "Inverses of rank invariant powers of a matrix," *J. Soc. Indust. Appl. Math. Numer. Anal.*, V, no. 1 (1968) 182–197.

Courant, T., and D. Hilbert
Methoden der Mathematischen Phys, Vol. I. Berlin: Springer, 1931.

Croisot, R.
"Demi-groups inversifs et demi-groups reunions de semi-groups simples," *Ann. Sci. École Norm. Sup.*, LXX (1953), 361–379.

Decell, H. P., Jr.
[1] "An alternate form for the generalized inverse of an arbitrary complex matrix," *SIAM Rev.*, VII (1965), 356–358.

[2] "An application of the Cayley-Hamilton theorem to generalized matrix inversion," *SIAM Rev.*, VII (1965), 526–528.

[3] "A characterization of the maximal subgroups of the semi-group of $n \times n$ complex matrices," to appear in *Czechoslovak Math. J.*

[4] "A special form of the generalized inverse of an arbitrary complex matrix," *NASA TN D–2784*, Washington, D. C., 1965.

[5] "An application of generalized matrix inversion to sequential least squares parameter estimation," *NASA TN D–2830*, Washington, D.C., 1965.

Decell, H. P., Jr., and S. W. Kahng
"An interative method for computing the generalized inverse of a matrix," *NASA TN D–3464*, Washington, D. C., June 1966.

Decell, H. P., Jr., and P. L. Odell
[1] "A note concerning a generalization of the Gauss-Markov theorem," *Texas Journal of Science*, XXVII, no. 1 (1966), 21–24.

[2] "On the fixed point probability vector of regular or ergodic transition matrices," *Jour. Amer. Statist. Assoc.*, LXII (1967).

Dekerlegand, R. J.
"Analysis of generalized inverse computation schemes," thesis, University of Southwestern Louisiana, Lafayette, La., 1967.

Delaney, F. C., G. G. Gaffney and F. M. Speed
"Efficiency of generalized matrix inversion methods," *NASA-MSC Internal Note*, MSC-IN-66-ED-41, Houston, Texas, September 1966.

Delaney, F. C., and F. M. Speed
"A new algorithm for calculating the generalized inverse of an arbitrary real $m \times m$ matrix," *NASA-MSC Internal Note*, MSC-IN-66-ED-42, Houston, Texas, September 1966.

Dent, B. A., and J. Newhouse
"Polynomials orthogonal over a discrete domain," *SIAM Rev.*, I (1959), 55–59.

Desoer, C. A., and B. H. Whalen
"A note on pseudoinverses," *J. Soc. Indust. Appl. Math.*, XI (1963), 442–447.

Doković, D.
"On the generalized inverse for matrices," *Glasnik Mat.-Fiz. Astron. Ser. II. Drustvo Mat. Fiz. Hrvatske*, XX (1965), 51–55.

Dommanget, J.

"L'inverse d'un cracovien rectangulaire: Son emploi dans la resolution des sustemes d'equations lineaires," *Publ. Sci. Tech. Ministére de l'Air (Paris), Notes Tech. No. 128* (1963), 11–41.

Douglas, J., Jr., and C. M. Pearcy

"On convergence of alternating direction procedures in the presence of singular operators," *Numer. Math.*, V (1963), 175–184.

Doust, A., and V. E. Price

"The latent roots and vectors of a singular matrix," *Comput. J.*, VII (1964), 222–227.

Dragomir, P.

[1] "On the Greville-Moore formula for calculating the generalized inverse matrix," *An. Univ. Timisoara Ser. Sti. Mat.-Fiz.*, no. 1 (1963), 115–119.

[2] "The generalized inverse of a bilinear form," *An. Univ. Timisoara Sti. Ser. Mat.-Fiz.*, no. 2 (1964), 71–76.

Drazin, M. P.

"Pseudo-inverses in associative rings and semi-groups," *Amer. Math. Monthly*, LXV (1958), 506–514.

Dück, Werner

"Einzelschrittverfahren zur matrizeninversion," *Z. Angew. Math. Mech.*, XLIV (1964), 401–403.

Dwyer, P. S., and M. S. Macphail

"Symbolic matrix derivatives," *Ann. Math. Statist.*, XIX (1948), 517–534.

Eckart, C., and G. Young

[1] "The approximation of one matrix by another of lower rank," *Psychometrika*, I (1936), 211–218.

[2] "A principal axis transformation for non-Hermitian matrices," *Bull. Amer. Math. Soc.*, XLV (1939), 118–121.

Egervary, E.

[1] "On rank diminishing operations and their applications to the solution of linear equations," *Z. Angew. Math. Phys.*, XI (1960), 376–386.

[2] "Uber eine kunstruktive methode zur reduktion einer matrix auf die Jorkansche normalform," *Acta. Math. Acad. Sci. Hungar*, X (1959), 31–54.

Elliott, W. W.

[1] "Generalized Green's functions for compatible differential systems," *Amer. J. Math.*, L (1928), 253–258.

[2] "Green's functions for differential systems containing a parameter," *Amer. J. Math.*, LI (1929), 397–416.

Englefield, M. J.

"The commuting inverses of a square matrix," *Proc. Cambridge Philos. Soc.*, LXII (1966), 667–671.

Erdelyi, I.

[1] "A generalized group-inverse of square matrices," unpublished report.

[2] "On partial isometries in finite-dimensional Euclidean spaces," *J. Soc. Indust. Appl. Math.*, XIV, no. 3 (1966), 453–467.

[3] "On the 'reverse order law' related to the generalized inverse of matrix products," *J. Assoc. Comput. Mach.*, XIII (1966), 439–443.

[4] "The quasi-commuting inverses for a square matrix," *Atti Accad. Naz. Lencei Rend. Cl. Sci. Fis. Mat. Natur.* (8), XLII (1967), 626–633.

[5] "On the matrix equation $Ax = \lambda Bx$," *J. Math. Anal. Appl.*, XVII, no. 1 (1967), 119–132.

[6] "On normal partial isometries in finite-dimensional Euclidean spaces." Manhattan, Kansas: Technical Report 6, Kansas State University, July 1967.

[7] "Partial isometries closed under multiplication on Hilbert spaces," *J. Math. Anal. Appl.*, in press.

[8] "Partial isometries defined by a spectral property on unitary spaces." Manhattan, Kansas: Technical Report 12, Kansas State University, The Department of Statistics, March 1968.

[9] "Normal partial isometries closed under multiplication on unitary spaces," *Atti Accad. Naz. Lincei Rend. Cl. Sci. Fis. Mat. Natur.* (8), XLIII (1968), in press.

Faddeev, D. K., and V. N. Faddeeva
Computational Methods of Linear Algebra. San Francisco, Cal.: W. H. Freeman, 1963, 260–265.

Faddeev, D. K., V. N. Kublanovskaya, and V. N. Faddeeva
"On linear algebraic systems with rectangular and ill conditioned matrices," *Colloque International de Mathematiques du CNRS Besancon* (September 1966).

Fan, Ky, and A. J. Hoffman
"Some metric inequalities in the space of matrices," *Proc. Amer. Math. Soc.*, VI (1955), 111–116.

Fisher, A. G.
"On construction and properties of the generalized inverse," *J. Soc. Indust. Appl. Math.*, XV (1967), 269–272.

Florentin, J. J.
"Optimal control of continuous time, Markov, stochastic systems," *J. Electronics Control*, X (1961), 473–488.

Foster, Manus
"An application of the Wiener-Kolmogorov smoothing theory to matrix inversion," *J. Soc. Indust. Appl. Math.*, IX (1961), 387–392.

Foulis, D. J.
"Relative inverses in Baer *-semigroups," *Michigan Math. J.*, X (1963), 65–84.

Frame, J. S.
"Matrix functions and applications. I. Matrix operations and generalized inverses," *IEEE Spectrum* (1964), 209–220.

Franck, P.
"Sur la distance minimale d'une matrice régulière donnée au lieu des matrices singulières," *Deux. Congr. Assoc. Francaise Calcul. et Traitement Information*, Paris: Gauthier-Villars, 1962, 55–60.

Fredholm, I.
"Sur une class d'"equations fonctionnelles," *Acta Math.*, XXVII (1903), 365–390.

Freidman, B.
Principles and Techniques of Applied Mathematics. New York: Wiley, 1956.

Friedrichs, K. O.
"Functional analysis and applications." New York: Lecture Notes, New York University, 1953.

Fulkerson, D. R., and O. A. Gross
"Incidence matrices and interval graphs," *Pacific J. Math.*, XV no. 3 (1965), 835–855.

Gabriel, R.
[1] "Estinderea complementilor algebrici generalizati la matrici oarecare," *Stud. Cerc. Mat.*, X (1965).
[2] "Uber die verallgemeinerte inverse einer matrix," unpublished manuscript, 1967.

Gaches, J., J. L. Rigal, and X. Rousset de Pina.
"Distance Euclidienne d'une application lineaire σ au lieu des applications de rang *r* donne. Determination d'une meilleure approximation de rang *r*," *C. R. Acad. Sci. Paris*, CCLX (1965), 5672–5674.

Gainer, P. A.
"A method for computing the effect of an additional observation on a previous least-square estimate," *NASA TN D-1599*, Washington, D. C., 1963.

Gantmacher, F. R.
[1] *Applications of the Theory of Matrices*, translated by J. L. Brenner. New York: Interscience, 1959
[2] *The Theory of Matrices, Vols. I and II*. New York: Chelsea House, 1959.

Gately, W. Y.
"Application of the generalized inverse concept to the theory of linear statistical models' Doctoral dissertation, Oklahoma State University, Stillwater, Okla., 1962.

Gavurin, M. K.
"Ill conditioned systems of linear algebraic equations," *Internat. J. Comput. Math.*, I (1964/65), 36–50.

Giurescu, C., and R. Gabriel
"Some properties of the generalized matrix inverse and semi-inverse," *An. Univ. Timisoara Ser. Sti Mat.-Fiz.*, no. 2 (1964), 103–111.

Glassey, C. R.
"An orthogonalization method of computing the generalized inverse of a matrix." Berkeley, Cal.: Report ORC 66–10, University of California, College of Engineering, Operations Research Center, April 1966.

Goldberger, A. S.
"Stepwise least squares residual analysis and specification error," *J. Amer. Statist. Assoc.*, LVI (December 1961), 998–1000.

Goldman, A. J., and M. Zelen
"Weak generalized inverses and minimum variance linear unbiased estimation," *J. Res. Nat. Bur. Standards Sect. B.*, LXVIII, B (1964), 151–172

Golub, G.
"Numerical methods for solving linear least squares problems," *Numer. Math.*, VII (1965), 206–216.

Golub, G., and W. Kahan
"Calculating the singular values and pseudoinverse of a matrix," *J. Soc. Indust. Appl. Math. Numer. Anal.*, II (1965), 205–224.

Good, I. J.
"On the independence of quadratic expression," *J. R. Statist. Soc. B*, XXV (1963), 377–382.

Graybill, F. A.

[1] *An Introduction to Linear Statistical Models*, I. New York: McGraw-Hill, 1961.

[2] *Introduction to Matrices with Applications in Statistics*. Belmont, Cal.: Wadsworth Publishing Company, 1969.

Graybill, F. A., and G. Marsaglia

"Idempotent matrices and quadratic forms in the general linear hypothesis," *Ann. Math. Statist.*, XXVIII (1957), 678–686.

Graybill, F. A., C. D. Meyer, and R. J. Painter

"Note on the computation of the generalized inverse of a matrix," *SIAM Rev.*, VIII, no. 4 (1966), 522–524.

Greub, W., and W. C. Reinboldt

"Non-self-adjoint boundary value problems in ordinary differential equations," *J. Res. Nat. Bur. Standards Sect. B*, LXIVB (1960), 83–90.

Greville, T. N. E.

[1] "E. H. Moore's generalization of the concept of inverse of a matrix to include rectangular and singular matrices," unpublished manuscript, circa 1956.

[2] "The pseudoinverse of a rectangular or singular matrix and its application to the solution of systems of linear equations," *SIAM Newsletter*, V (1957), 3–6.

[3] "On smoothing a finite table: A matrix approach," *J. Soc. Indust. Appl. Math.*, V (1957), 137–154.

[4] "The pseudoinverse of a rectangular matrix and its applications to the solution of systems of linear equations," *SIAM Rev.*, I (1959), 38–43.

[5] "Some applications of the pseudoinverse of a matrix," *SIAM Rev.*, II (1960), 15–22.

[6] "Note on fitting of functions of several independent variables," *J. Soc. Indust. Appl. Math.*, IX (1961), 109–115. Erratum: *ibid.*, 317.

[7] "Notes on matrix pseudoinverses." Bethlehem, Pa.: Lecture Notes, Lehigh University, Summer 1962.

[8] "Further remarks on the pseudoinverse of a matrix," unpublished manuscript, presented to Mathematics Club, University of Michigan, Ann Arbor, Mich., December 1962.

[9] "A product characterization of the generalized inverse of a singular square matrix." Madison, Wis.: The University of Wisconsin, July 1963. [Abstract in *Notices Amer. Math. Soc.*, X (1963), 472.]

[10] "Note on the generalized inverse of a matrix product," *SIAM Rev.*, VIII, no. 4 (1966), 518–521. Erratum: *SIAM Rev.*, IX (1967), 249.

[11] "Spectral generalized inverses of singular square matrices," Abstract in *Notices Amer. Math. Soc.*, XV (1968), 111.

[12] "Spectral generalized inverses of square matrices," *MRC Technical Summary Report* no. 823. Madison, Wis.: Mathematics Research Center, University of Wisconsin.

[13] "Generalized inverses of finite matrices," to be submitted to *SIAM Rev.*

Guedj, R.

"L'utilisation d'inverse généralisés dans la résolution de systèmes linéaires de rang guelconque," *Troisième Congr. de Calcul et de Traitement de l'Information AFCALTI* (1965), 137–143.

Gura, I. A.

"Notes on the pseudoinverse of a matrix," unpublished manuscript, 1967.

Halmos, P. R.
Finite-Dimensional Vector Spaces, 2nd ed. Princeton, N. J.: D. Van Nostrand, 1958.

Hamburger, H.
"Non-symmetric operators in Hilbert space," *Proceedings of the Symposium on Spectral Theory and Differential Problems*, 67–112. Stillwater, Okla.: Oklahoma Agricultural and Mechanical College, 1951.

Hamburger, H. L., and M. E. Grimshaw
Linear Transformations in n-Dimensional Vector Space. Cambridge, Mass.: Cambridge University Press, 1951.

Harada, S.
"An existence proof of the generalized Green's function," *Osaka Math. J.*, V (1953), 59–63.

Harris, W. A., Jr., and T. N. Helvig
[1] "Applications of the pseudoinverse to modeling," *Technometrics*, VIII, no. 2 (1966), 351–357.

[2] "Marginal and conditional distributions of singular distributions," *Publications of the Res. Inst. for Math. Sci.*, *Kyoto Univ.*, *Ser. A*, I (1966), 199–204.

Hawkins, J. B., and A. Ben-Israel
"On generalized matrix functions." Evanston, Ill.: Systems Research Memo. no. 193, Northwestern University, January 1968.

Hearson, J. Z.
[1] "Construction of EPr generalized inverses by inversion of nonsingular matrices," *J. Res. Nat. Bur. Standards Sec. B.*, LXXI (1967), 57–60.

[2] "A generalized matrix version of Rennie's inequality," *J. Res. Nat. Bur. Standards Sec. B*, LXXI (1967), 61–64.

[3] "Partially isometric matrices," *J. Res. Nat. Bur. Standards Sec. B.*, LXXI (1967), 225–228.

[4] "Symmetrizable generalized inverses of symmetrizable matrices," *J. Res. Nat. Bur. Standards Sec. B*, LXXI (1967), 229–231.

[5] "Polar factorization of a matrix," *J. Res. Nat. Bur. Standards Sec. B*, LXXI (1967), 65–67.

[6] "On the singularity of a certain bordered matrix," *J. Soc. Indust. Appl. Math.*, XV, no. 6 (November 1968), 1413–1421.

Hearon, J. Z., and J. W. Evans
[1] "Differentiable generalized inverses," submitted for publication. Bethesda, Md.: Mathematical Research Branch, NIAMD, National Institute of Health.

[2] "On spaces and maps of generalized inverses," to be published. Bethesda, Md.: Mathematical Research Branch, NIAMD, National Institute of Health.

Herring, G. P.
"A note on generalized interpolation and the pseudoinverse," *J. SIAM Numer. Anal.*, IV, no. 4 (1967), 548–556.

Hestenes, M. R.
[1] "Inversion of matrices by biorthogonalization and related topics," *J. Soc. Indust. Appl. Math.*, VI (1958), 51–90.

[2] "Relative self-adjoint operators in Hilbert space," *Pacific J. Math.*, XI (1961), 1315–1357.

[3] "Relative Hermitian matrices," *Pacific J. Math.*, XI (1961), 225–245.

[4] "A ternary algebra with applications to matrices and linear transformations," *Arch. National Mech. Anal.*, XI (1962), 138–194.

Hilbert, D.
"Gründzuge einer allgemeinen theorie der linearen integralgleichunben," *Teubner, Leipzig and Berlin* (1912). Reprint of six articles that appeared originally in the *Gottingen Nachrichten* (1904), pp. 49051; (1904), pp. 213–259; (1905), pp. 307–338; (1906), pp. 157–227; (1906), pp. 439–480; (1910), pp. 355–417.

Holder, E.
"Die Lichtensteinsche methods für die Entwicklung des sweiten variation, angewandt auf das problem von Lagrange," *Prace Mathematycano-Fizyczne*, XLIII (1935), 307–346.

Horst, P.
Matrix Algebra for Social Scientists. New York: Holt, Rinehart and Winston, 1963, Chapters 17–20.

Householder, A. S.
[1] *The Theory of Matrices for Numerical Analysis.* New York: Blaisdell, 1964.

[2] "Unitary triangularization of a nonsymmetric matrix," *J. Assoc. Comput. Mach.*, V (1958), 339–342.

Householder, A. S., and G. Young
"Matrix approximation and latent roots," *Amer. Math. Monthly*, XLV (1938), 165–171.

Hurwitz, W. A.
"On the pseudo-resolvent to the kernel of an integral equation," *Transactions, Amer. Math. Soc.*, XIII (1912), 405–418.

Ijiri, Yuju
"On the generalized inverse of an incidence matrix," *J. Soc. Indust. Appl. Math.*, XIII (1965), 827–836.

Jameson, A.
"Solution of the equation $AX + XB = C$ by immersion of an $M \times M$ or $N \times N$ matrix," *J. Soc. Indust. Appl. Math.*, XVI, no. 5 (1968), 1020–1023.

John, Peter W. M.
"Pseudo-inverses in the analysis of variance," *Ann. Math. Statist.*, XXXV (1967), 895–896.

Jones, John, Jr.
"On the Lyapunov stability criteria," *J. Soc. Indust. Appl. Math.*, XIII (1965), 941–945.

Kalman, R. E.
[1] "Contributions to the theory of optimal control," *Bol. Soc. Mat. Mexicana*, V (1960), 102–119.

[2] "A new approach to linear filtering and prediction problems," *Jour. of Basic Engineering* (March 1960), 35–44.

[3] "New methods and results in linear prediction and filtering theory," *Technical Report*, no. 61–1. Baltimore, Md.: RIAS, 1961.

Kalman, R. E., Y. C. Ho, and K. S. Narenda
"Controllability of linear dynamic systems," *Contributions to Differential Equations*, Vol. I. New York: Interscience, 1963, 189–213.

Kato, T.
"Perturbation theory for linear operators," *Die Grundlehren der Mathematischen Wissinschaften*, CXXXII. Verlin-Heidleverb-New York: Springer-Verlag, 1966.

Katz, I. J.
[1] "On the generalized inverse of a product and EPr matrices," to be published.
[2] "Weigmann type theorems for EPr matrices," *Duke Math. J.*, XXXII (1965), 423–427.

Katz, I. J., and M. H. Pearl
"On EPr and normal matrices," *J. Res. Nat. Bur. Standards Sect. B*, LXX (1966), 47–77.

Keller, H. B.
"On the solution of singular and semidefinite linear systems by iteration," *J. SIAM Numer. Anal.*, II (1965), 281–290.

Kellog, R. B., and J. Spanier
"On optimal alternating, direction parameters for singular matrices," *Math. Comp* XIX (1965), 448–452.

Kim, Jin Bai
"On singular matrices," *J. Korean Math. Soc.*, III (1966), 1–2.

Kirby, M. J. L.
"Generalized inverses and chance-constrained programming," Doctoral dissertation, Northwestern University, Evanston Ill., June 1965.

Koop, J. C.
[1] "Generalized inverse of a singular matrix," *Nature*, CXCVIII (June 1963), 1019–1020.
[2] "Generalized inverse of a singular matrix," *Nature*, CC (November 1963), 716.

Korganoff, A.
[1] "Functions of a normed vector space applied to the iterative solution of rectangular and square matrix non-linear equations of any given form," *Nordisk Symposium*, Oslo, August 18–22, 1961.
[2] "Les polynômes d'interpolation de matrices carrées a coefficients matriciels et les methodes iteratives de résolution numerique des equations de matrices carrées de forme quelconque," *Proc. IFIP Congress*, 1962, 102–106.
[3] "The inversion of rectangular matrices in the resolution of ill-condition linear systems," *Proc. Nordsam Congress*, Helsinki, August 16–20, 1963, Helsinki, 1964, Vol. 2, 179–190.

Korganoff, A., and M. Pavel-Parvu
[1] "Intreprétation a l'aide des pseudoinverses de la solution d'equations matricielles lineaires provenant de la discretisation d'operateurs differentiels et integraux," *83e Congres de l'Association Francaise pour l'avancement des Sciences*, Lille (July 1964).
[2] *Éléments de Théorie des Matrices Carrées et Rectangles en Analyse Numérique.* Paris: Dunod, 1967.
[3] *Méthodes de Calcul Numérique—2. Éléments de Théorie des Matrices Carrées et Rectangular en Analyse Numérique.* Paris: Dunod, 1967.

Kublanovskaya, V. N.
"On the computation of the generalized matrix inverse and the projection," *USSR J. Comp. Math. Phys.*, VI (1966), 326–332.

Kuo, M. C. Y., and L. F. Kazda
"Minimum energy problems in Hilbert function space," *J. Franklin Inst.*, CCLXXXIII (1967), 38–54.

Lanczos, C.
[1] *Linear Differential Operators*. Princeton, N. J.: D. Van Nostrand, 1961.
[2] "Linear systems in self-adjoint form," *Amer. Math. Monthly*, LXV (1958), 665–679.

Langenhop, C. E.
"On generalized inverses of matrices," *J. Soc. Indust. Appl. Math.*, XV (1967), 1239–1246.

Lass, J., and C. B. Solloway
"A note on the secular equation of the product of two matrices," *Amer. Math. Monthly*, LXVIII (1961), 906–907.

Lewis, D. C.
"On the role of first integrals in the perturbation of periodic solutions," *Ann. of Math.*, LXIII (1963), 535–548.

Lewis, T. O.
"Application of the theory of generalized matrix inversion to statistics." Doctoral dissertation, University of Texas, Austin, Texas, 1966.

Lewis, T. O., and T. Newman
"Pseudo-inverses of positive semi-definite matrices," *J. Soc. Indust. Appl. Math.*, XVI, no. 4 (December 1968), 701–703.

Lewis, T. O. and P. L. Odell
[1] "A generalization of the Gauss-Markov theorem," *J. Amer. Statist. Assoc.*, LXI (1967), 1063–1066.
[2] *Estimation in Linear Models*. Englewood, N.J.: Prentice-Hall, 1971.

Liber, A. E.
"On the theory of generalized groups," *Dokl. Akad. Nauk SSSR (N.S.)*, XCVII 1954), 25–28.

Ljapin, E. S.
[1] "Inversion of elements in semigroups," *Leningrad Gos. Ped. Inst. Ucen. Zap.*, CLXVI (1958), 65–74.
[2] "Semigroups, Translations of Math. Monographs, *Amer. Math. Soc.*, III (1963), 447.

Loud, W. S.
"Generalized inverses and generalized Green's functions," *J. Soc. Indust. Appl. Math.*, XIV, no. 2 (1966), 342–369.

MacDuffee, C. C.
The Theory of Matrices. New York: Chelsea House, 1956, pp. 110.

Magness, T. A., and T. B. McGuire
"Comparison of least squares and minimum variance estimates on regression parameters," *Ann. Math. Statist.*, XXXIII (1962).

Malik, H. J.
"A note on generalized inverses," *Naval Research Logistics Quarterly*, XV, no. 4 (1968), 605–612.

Marcus, M., and H. Minc
   A Survey of Matrix Theory and Matrix Inequalities. Boston, Mass.: Allyn and Bacon, 1964, pp. 110.

Marsaglia, G.
   "Conditional means and variances of normal variables with singular covariance matrix," *J. Amer. Statist. Assoc.* (1964), 1203–1204.

Mayne, D.
   "An algorithm for the calculation of the pseudoinverse of a singular matrix," *Comput. J.*, IX (1966), 312–317.

Meicler, M.
   [1] "Weighted generalized inverses with minimal p and q norms," Doctoral dissertation, University of Texas, Austin, Texas, 1966.

   [2] "Chebyshev solution of an inconsistent system of $n + 1$ linear equations in n unknowns in terms of its least squares solution," *SIAM Rev..*, X (July 1968), 373–375.

Meicler, M., and P. L. Odell
   [1] "Weighted generalized inverses," *NASA NAS* 9–5384 (February 1967), 1–13.

   [2] "Report: $p - q$ generalized inverses," *NASA NAS 9–5384* (January 1967), 1–19.

Meyer, Carl D.
   [1] "On ranks of pseudoinverses," *SIAM Rev.*, XI, no. 3 (1969), 382–386.

   [2] "On the rank of the sum of two rectangular matrices," *Canadian Math. Bull.*, XII, no. 4 (1969), 508.

   [3] "On the construction of solutions to the matrix equations $AX = A$ and $YA = A$," *SIAM Rev.*, XV, no. 4, (1969), to appear.

   [4] "Some remarks on EPr matrices and generalized inverses," *Linear Algebra and its Appl.*, III, no. 1 (1970), to appear.

   [5] "Generalized inverses of triangular matrices," *SIAM J. Appl. Math.*, XVIII, no. 2 (1970), to appear.

Meyer, C. D., and R. J. Painter
   "Note on a least squares inverse for a matrix," *J. Assoc. Comp. Mach.*, XVII, no. 1 (1970), 110–112.

Milne, R. D.
   "An oblique matrix pseudoinverse," *J. Soc. Indust. Appl. Math.*, XVI, no. 5 (1968), 931–944.

Mitra, S. K.
   [1] "A new class of g-inverses of square matrices," *Sankhya*, Series A, XXX, no. 3 (1968), 323–331.

   [2] "On a generalized inverse of a matrix and applications," *Sankhya*, Series A, XXX, no. 1 (1968), 107–114.

   [3] "The subalgebra generated by a square matrix and generalized inverse." Calcutta, India: Tech. Report #11/68, Research and Training School, Indian Statist. Inst., April 30, 1968.

Mitra, S. K., and C. R. Rao
   "Simultaneous reduction of a pair of quadratic forms," *Sankhya*, Series A, XXX, no. 3 (1968), 313–322.

Mizel, V. J., and M. M. Rao
   "Non-symmetric projections in Hilbert space," *Pacific J. Math.*, XII (1962), 343–357.

Moore, E. H.

[1] Abstract in *Bull. Amer. Math. Soc.*, XXVI (1920), 394–395.

[2] "Generalized analysis, part I," *Mem. Amer. Philos. Soc.*, I (1935).

Morris, G. L.

"Characterizations of generalized inverses for matrices," Doctoral dissertation, Texas Tech. College, Lubbock, Texas, 1967.

Morris, G. L., and P. L. Odell

[1] "A characterization for generalized inverses of matrices," *SIAM Rev.*, X (1968), 208–211.

[2] "Common solutions for *n* matrix equations with applications," *J. Assoc. Comput. Mach.*, XV, no. 2, (1968), 272–274.

Morrison, D. D.

"Remarks on the unitary triangularization of a nonsymmetric matrix," *J. Assoc. Comput. Mach.*, VII (1960), 185–195.

Munn, W. D.

"Pseudoinverse in semi-groups," *Proc. Cambridge Philos. Soc.*, LVII (1961), 247–250

Munn, W. D., and R. Penrose

"A note on inverse semi-groups," *Proc. Cambridge Philos. Soc.* (England), L (1955), 396–399.

Murray, F. J., and J. von Neumann

"On rings of operators—I," *Ann. of Math.*, XXVII (1936), 116–229.

Myller, A.

"Gewohnliche Differentialgleichungen Hoherer Ordnung in ihre Beziehung zu den Integralgleichungen." Inaugural-dissertation, Gottingen, 1906.

Nelson, D. L.

"Numerical methods for the solution of non-linear least squares problems," Doctoral dissertation, Texas Tech University, Lubbock, Texas, 1969.

Newman, T. G., and P. L. Odell

"On the concept of a $p - q$ generalized inverse of a matrix," *SIAM J. Appl. Math.*, XVII, no. 3 (1969), 520–525.

Nobel, B.

"A method for computing the generalized inverse of a matrix," *J. Soc. Indust. Appl. Math. Numer. Anal.*, III, no. 4 (1966), 582–584.

Odell, P. L., and H. P. Decell, Jr.

"On computing the fixed point probability vector of regular or ergodic transition matrices," *J. Assoc. Comput. Mach.*, XIV, no. 4 (1967), 765–768.

Orey, S.

"Potential kernels for recurrent Markov chains," *J. Math. Anal. Appl.*, VIII (1964), 104–132.

Osborne, E. E.

[1] "On least square solutions of linear equations," *J. Assoc. Comput. Mach.*, VIII (1961), 628–636.

[2] "Smallest least squares solutions of linear equations," *J. Soc. Indust. Appl. Math. Numer. Anal.*, II (1965), 300–307.

Paige, L. J., and J. D. Swift

Elements of Linear Algebra. Boston, Mass.: Ginn and Company, 1961.

Pearl, M. H.
   [1] "On normal and EPr matrices," *Michigan Math. J.*, VI (1959), 1–5.
   [2] "On normal and EPr matrices," *Michigan Math. J.*, VI (1959), 89–94.
   [3] "On generalized inverses of matrices," *Proc. Cambridge Philos. Soc.*, LXII (1966), 673–677.

Penrose, R.
   [1] "A generalized inverse for matrices," *Proc. Cambridge Philos. Soc.* (England), LI (1955), 406–413.
   [2] "On best approximate solutions of linear matrix equations," *Proc. Cambridge Philos. Soc.* (England), LII (1956), 17–19.

Pereyra, V., and J. B. Rosen
   "Computation of the pseudoinverse of a matrix of unknown rank." Stanford Cal.: Tech. Rep. CS 13, Stanford University, Computer Science Div., September 1964.

Perlis, S.
   Theory of Matrices. Cambridge, Mass.: Addison-Wesley, 1958.

Petryshyn, W. V.
   [1] "On the inversion of matrices and linear operators," *Proc. Amer. Math. Soc.*, XVI (1965), 893–901.
   [2] "On generalized inverse and on the uniform convergence of $(I - \beta\kappa)^n$ with application to iterative methods," *J. Math. Anal. Appl.*, XVIII (1967), 417–439.

Popa, C-Zinljh
   "Note on the inversion of singular matrices," *Lucrar Sti. Inst. Ped. Timisoara Mat.-Fiz.* (1962), 149–153.

Porter, W. A.
   Modern Foundation of System Engineering. New York: Macmillan, 1966, pp. 493.

Presic, S. B.
   "Certaines equations matricielles," *Publications de la faculte d'electro-technique de l'universite a Belgrade Serie: Math. et Phys.*, no. 121 (1963), 31–32.

Preston, G. B.
   "Inverse semi-groups," *J. London Math. Soc.*, XXIX (1954), 396–403.

Price, C. M.
   "The matrix pseudoinverse and minimal variance estimates," *SIAM Rev.*, VI (1964), 115–120.

Pyle, L. D.
   [1] "The generalized inverse in linear programming," Doctoral dissertation, Purdue University, Lafayette, Ind., 1960.
   [2] "The generalized inverse in linear programming—basic theory," *Technical Report* no. 1, Lafayette, Ind.: Purdue University, Computer Science Dept., August 1, 1965.
   [3] "Generalized inverse computations using the gradient projection method," *J. Assoc. Comput. Mach.*, XI (1964), 422–429.
   [4] "A generalized inverse ε-algorithm, for construction intersection projection matrices, with application," *Numer. Math.*, X (1967), 86–102.

Rado, R.
   "Note on generalized inverses of matrices," *Proc. Cambridge Philos. Soc.* (England), LII (1956), 600–601.

Rainbolt, M. B.

"Sequential least squares parameter estimation," NASA-MSC unpublished report.

Rao, C. R.

[1] "Analysis of dispersion for multiply classified data with unequal numbers in cells," *Sankhya*, XV (1955), 253–280.

[2] "A note on a generalized inverse of a matrix with applications to problems of mathematical statistics," *J. Roy. Statist. Soc. B* (England), XXIV (1962), 152–158.

[3] *Linear Statistical Inference and its Applications.* New York: Wiley, 1965, pp. 522.

[4] "Generalized inverse for matrices and its applications in mathematical statistics," *Festschrift for J. Neyman: Research Papers in Statistics*, London: Wiley, 1966, 263–279.

[5] "Calculus of generalized inverses of matrices. Part I: General theory," *Sankhya Ser. A*, XXIX (1967), 317–342.

Rao, K. K.

"A simplified proof of Gauss-Markov theorem when the regression matrix is of less than full rank," *Amer. Math. Monthly*, LXXII (1966), 394–395.

Rayner, A. A., and D. Livingstone

"On the distribution of quadratic forms in singular normal variates," *South African J. Agricultural Sci.*, VIII (1965), 357–370.

Rayner, A. A., and R. M. Pringle

"A note on generalized inverses in the linear hypothesis not of full rank," *Ann. Math. Statist.*, XXXVIII, no. 1 (1967), 271–274.

Reid, W. T.

[1] "Generalized Green's matrices for compatible systems of differential equations," *Amer. J. Math.*, LIII (1931), 443–459.

[2] "Reccati matrix differential equations and non-oscillation criteria for associated linear differential systems," *Pacific J. Math.*, XIII (1963), 665–685.

[3] "Principal solutions of non-oscillatory linear differential systems," *J. Math. Anal. Appl.*, IX (1964), 397–423.

[4] "A matrix equation related to a non-oscillation criterion and Liápunov stability," *Quart. Appl. Math.*, XXIII (1965), 83–87.

[5] "Generalized Green's matrices for two-point boundary problems," *J. Soc. Indust. Appl. Math.*, XV (1967), 856–870.

Robinson, D. W.

"On the generalized inverse of an arbitrary linear transformation," *Amer. Math. Monthly*, LXIX (1962), 412–416.

Rohde, C. A.

[1] "Contributions to the theory, computation and applications of generalized inverses," Doctoral dissertation, North Carolina State University, Chapel Hill, N.C., 1964).

[2] "Generalized inverses of partitioned matrices," *J. Soc. Indust. Appl. Math.*, XIII (1965), 1033–1035.

[3] "Some results on generalized inverses," *SIAM Rev.*, VIII, no. 2 (1966), 201–205.

Rohde, C. A., and J. R. Harvey

"Unified least squares analysis," *J. Amer. Statist. Assoc.*, LX (1965), 523–527.

Rosen, J. B.
[1] "The gradient projection method for nonlinear programming, part I: Linear constraints," *J. Soc. Indust. Appl. Math.*, VIII (1960), 181–217.

[2] "The gradient projection method for nonlinear programming, part II: Nonlinear constraints," *J. Soc. Indust. Appl. Math.*, IX (1961), 514–532.

[3] "Minimum and basic solutions to singular linear systems," *J. Soc. Indust. Appl. Math.*, XII (1964), 156–162.

Rust, B., W. R. Burrus, and C. Schneeberger
"A simple algorithm for computing the generalized inverse of a matrix," *Comm. ACM*, IX, no. 5 (1966), 381–386.

Schneider, H.
"A matrix problem concerning projections," *Proc. Edinburgh Math. Soc.*, X, no. 2 (1956), 129–130.

Schneider, H., and G. P. Barker
Matrices and Linear Algebra. New York: Holt, Rinehart and Winston, 1968.

Schulz, G.
"Iterative berechnung der reziproken matrix," *Z. Angew. Math. Mech.*, XIII (1933), 57–59.

Schwerdtfeger, H.
Introduction to Linear Algebra and the Theory of Matrices. Groningen, P. Noordhoff, 1961.

Scroggs, J. E., and P. L. Odell
"An alternate definition of a pseudo-inverse of a matrix," *J. Soc. Indust. Appl. Math.*, XIV (1966), 796–810.

Searle, S. R.
[1] "Additional results concerning estimable functions and generalized inverse matrices," *J. Roy. Statist. Soc. Ser. B*, XXVII (1965), 486–490.

[2] *Matrix Algebra for the Biological Sciences.* New York: Wiley, 1966, Chapters 6, 9, and 10.

Seidel, J. J.
"Angles and distances in n-dimensional euclidean and noneuclidean geometry, I, II, III," *Nedrl. Akad. Wetensch. Proc. Ser. A*, XVII (1955), 329–335, 336–340, 535–541.

Shanbhag, D. N.
"On the independence of quadratic forms," *J. R. Statist. Soc. B*, XVIII (1966), 582–583.

Sharpe, G. E., and G. P. H. Styan
[1] "A note on the general network inverse," *IEEE Trans. Circuit Theory*, XII (1965), 632–633.

[2] "Circuit duality and the general network inverse," *IEEE Ttans. Circuit Theory*, XII (1965), 22–27.

[3] "A note on equicofactor matrices," *Proc. IEEE*, LV, no. 7 (1967), 1226–1227.

Sheffield, R. D.
[1] "On pseudo-inverses of linear transformations in Banach spaces." Oak Ridge, Tenn.: Oak Ridge Nat. Lab. Report, no. 2133 (1956).

[2] "A general theory of linear systems," *Amer. Math. Monthly*, LXV (1958), 109–111.

Showalter, D.
"Representation and computation of the pseudoinverse," *Proc. Amer. Math. Soc.*, XVIII (1967), 584–587.

Shurbet, G. L., T. O. Lewis, and H. W. Milnes
"Recovery of linear transformations using collinear invariant points and pseudo-inverses," *Texas Journal of Science*, XX, no. 4 (April 1969), 361–366.

Skornyakov, L. A.
Complemented Modular Lattices and Regular Rings. Edinburgh-London: Oliver and Boyd, 1964, pp. 182.

Speed, F. M., and A. Feiveson
"Testing hypotheses and generalized inverse," unpublished report.

Stewart, G. W., III
[1] *Perturbation Bounds for the Linear Least Squares Problem*. Oak Ridge, Tenn.: Computing Technological Center, 1966.
[2] "On the continuity of the generalized inverse," *J. Soc. Indust. Appl. Math.*, XVII (January 1969), 33–45.

Tapley, B. D., and P. L. Odell
"A study of optimum method for determining and predicting space vehicle trajectories and control programs," *Texas Center for Research Memo.*, no. 1 (April 15, 1965), 43–62.

Taylor, A. E.
Introduction to Functional Analysis. New York: Wiley, 1964.

Taylor, A. E., and C. J. A. Halberg, Jr.
"General theorems about a bounded linear operator and its conjugate," *J. Reine Angew. Math.*, CXCVIII (1957), 93–111.

Tewarson, R. P.
[1] "A direct method for generalized matrix inversion," *J. Soc. Indust. Appl. Math. Numer. Anal.*, IV, no. 4 (1967), 499–507.
[2] "On some representations of generalized inverses," *SIAM Rev.*, XI (April 1969), 272–276.

Tseng, Y.
[1] "The characteristic value problem of Hermitian functional operators in a non-Hilbertian space," Doctoral dissertation, University of Chicago, Dept. of Math., Chicago, Ill., 1933.
[2] "Sur les solutions des equations operatrices functionnelles entre les espaces unitaires," *C. R. Acad. Sci. Paris*, CCXXVIII (1949), 640–641.
[3] "Generalized inverses of unbounded operators between two unitary spaces," *Dokl. Akad. Nauk. SSSR (N.S.)*, LXVII (1949), 431–434. Reviewed in *Math. Rev.*, XI (1950), 115.
[4] "Properties and classifications of generalized inverses of closed operators," *Dokl., Akad. Nauk. SSSR (N.S.)*, LXVII (1949), 607–610. Reviewed in *Math. Rev.*, XI (1950), 115.
[5] "Virtual solutions and general inversions," *Uspehi. Mat. Nauk. (N.S.)*, XI (1956), 213–215. Reviewed in *Math. Rev.*,

Urquhart, N. S.
[1] "Computation of generalized inverse matrices which satisfy specified conditions," *SIAM Rev.*, X (1968), 216–218.

[2] "The nature of the back of uniqueness of generalized inverse matrices," *SIAM Rev.*, XI (April 1969), 268–271.

Vagner, V. V.
"Generalized groups," *Dokl. Akad. Nauk. SSSR*, LXXXIV (1952), 1119–1122.

Van der Vaart, H. R.
Appendix to "Generalization of Wilcoxen statistic for the case of *k* samples," by E. H. Yen, *Statistica Neerlandica*, XVIII (1964), 303–305.

Von Neumann, J.
[1] "Uber adjungierte funktionaloperation," *Ann. of Math.*, XXXIII (1932), 294–310.

[2] "On regular rings," *Proc. Nat. Acad. Sci. USA*, XXII (1936), 707–713.

[3] "Functional operators, vol. II. The geometry of orthogonal spaces," *Ann. of Math. Studies No. 29*. Princeton, N. J.: Princeton University Press, 1950.

[4] *Continuous Geometry*. Princeton, N. J: Princeton University Press, 1960, pp. 299.

Vortuba, G.
"Generalized inverses and singular equations in functional analysis," Doctoral dissertation, University of Michigan, Dept. of Math., Ann Arbor, Mich., 1963.

Wedderburn, J. H. M.
"Lectures on matrices." Providence, R. I.: American Math. Soc., 1934.

Wedin, Per-Ake
"Results on pseudoinverses," unpublished manuscript, 1967.

Westfall, W. D. A.
[1] "Existence of the generalized Green's function," *Ann. of Math.*, X, no. 2 (1909), 177–181.

[2] "Zur tjeorie der integralgleichungen." Inaugural-dissertation, Gottingen, 1905.

Wilkinson, J. H.
[1] "Calculation of the eigenvalues of a symmetric tridiagonal matrix by the method of disection," *Numer. Math.*, IV (1962), 362–367.

[2] "The calculation of the eigenvectors of co-diagonal matrices," *Comput. J.*, I (1958), 148–152.

Willner, L. B.
"An elimination method for computing the generalized inverse," *Math. Comp.*, **21**, no. 98 (1967), 227–229.

Wyler, O.
[1] "Green's operators," *Ann. Mat. Pura Appl.*, LXVI, no. 4 (1964), 251–264.

[2] "On two-point boundary problems," *Ann. Mat. Pura Appl.*, LXVII, no. 4 (1965), 127–142.

Zacks, S.
"Generalized least squares estimators for randomized fractional replication designs," *Ann. Math. Statist.*, XXXV (1964), 696–704.

Zadeh, L. A., and C. A. Desoer
Linear System Theory. New York: McGraw-Hill, 1963, Chapter 17.

Zimmule, D.
"Techniques for computing the pseudo-inverse of a matrix," unpublished manuscript.

Zelen, M.
"The role of constraints in the theory of least squares." Madison, Wis.: MRC Technical Summary Report no. 312, University of Wisconsin, Math. Research Center, U. S. Army, 1962.

INDEX

DATE DUE

GAYLORD			PRINTED IN U.S.A.